FIELD GUIDE TO THE SEASHORES OF SOUTH-EASTERN AUSTRALIA

CHRISTINE PORTER, TY G. MATTHEWS,
ALECIA BELLGROVE AND GEOFF WESCOTT

Copyright The Authors 2023. All rights reserved.

Except under the conditions described in the *Australian Copyright Act 1968* and subsequent amendments, no part of this publication may be reproduced, stored in a retrieval system or transmitted in any form or by any means, electronic, mechanical, photocopying, recording, duplicating or otherwise, without the prior permission of the copyright owner. Contact CSIRO Publishing for all permission requests.

The authors assert their right to be known as the creators of this work.

A catalogue record for this book is available from the National Library of Australia.

ISBN: 9781486315123 (pbk)
ISBN: 9781486315130 (epdf)
ISBN: 9781486315147 (epub)

How to cite:
Porter C, Matthews TG, Bellgrove A, Wescott G (2023) *Field Guide to the Seashores of South-Eastern Australia*. CSIRO Publishing, Melbourne.

Published by:

CSIRO Publishing
Private Bag 10
Clayton South VIC 3169
Australia

Telephone: +61 3 9545 8400
Email: publishing.sales@csiro.au
Website: www.publish.csiro.au
Sign up to our email alerts: publish.csiro.au/earlyalert

Front cover: (top) Thirteenth Beach reef (photo by C. Porter; (bottom, left to right) *Nectria ocellata* (photo by C. Porter), *Chaetomorpha coliformis* (photo by D. Squire), *Mictyris longicarpus* (photo by T. Matthews).
Back cover: (left to right) *Durvillaea potatorum* (photo by C. Porter), *Hydatina physis* (photo by T. Matthews), *Holopneustes porosissimus* (photo by C. Porter).

Edited by Natalie Korszniak
Cover design by Cath Pirret
Typeset by Envisage Information Technology
Printed in China by 1010 Printing International Ltd

CSIRO Publishing publishes and distributes scientific, technical and health science books, magazines and journals from Australia to a worldwide audience and conducts these activities autonomously from the research activities of the Commonwealth Scientific and Industrial Research Organisation (CSIRO). The views expressed in this publication are those of the author(s) and do not necessarily represent those of, and should not be attributed to, the publisher or CSIRO. The copyright owner shall not be liable for technical or other errors or omissions contained herein. The reader/user accepts all risks and responsibility for losses, damages, costs and other consequences resulting directly or indirectly from using this information.

CSIRO acknowledges the Traditional Owners of the lands that we live and work on across Australia and pays its respect to Elders past and present. CSIRO recognises that Aboriginal and Torres Strait Islander peoples have made and will continue to make extraordinary contributions to all aspects of Australian life including culture, economy and science. CSIRO is committed to reconciliation and demonstrating respect for Indigenous knowledge and science. The use of Western science in this publication should not be interpreted as diminishing the knowledge of plants, animals and environment from Indigenous ecological knowledge systems.

The paper this book is printed on is in accordance with the standards of the Forest Stewardship Council® and other controlled material. The FSC® promotes environmentally responsible, socially beneficial and economically viable management of the world's forests.

Jan23_01

Contents

Preface and acknowledgements   v
Start here to get the most out of this guide   vii

Introduction   1
    Conservation Code   1
    Preparing for your visit to the shoreline   2
    About scientific names   2
    Sea Country and Saltwater Peoples   5
    Human impacts on seashores in south-eastern Australia   6
    Conservation and management of south-eastern Australian shores   8
    Citizen science (by Rebecca Koss)   10
    Intertidal ecology   11

Plants   16
    Introduction to marine plants   16
    Blue–green algae   18
    Red algae   19
    Green algae   25
    Brown algae   34
    Seagrasses   49
    Saltmarsh plants   53
    Grey mangrove   57
    Lichens   58

Animals of rocky shores   59
    Introduction   59
    Sponges   63
    Anemones   64
    Flatworms   69
    Segmented or bristle worms   70
    Ribbon worms   77

Peanut worms	78
Shells	79
Chitons	79
Snails	82
Sea slugs and nudibranchs	108
Bivalves	110
Octopus	114
Animals with jointed limbs	116
Barnacles	116
Slaters, shrimp and crabs	123
Lace coral	143
Animals with tube feet	144
Urchins	144
Sea stars	146
Brittle stars	153
Sea cucumbers	154
Sea squirts, tunicates	155
Animals of sandy and muddy shores	**158**
Important marine soft-sediment habitats	158
Life beneath your beach towel	161
Typical beach bugs	162
Sandy beach predators	162
Like a bug stuck in the mud	163
Segmented or bristle worms	165
Shells	168
Molluscs with coiled shells	168
Molluscs with two shells	175
Animals with jointed limbs	181
Sand hoppers, sea slaters, shrimp and crabs	181
Millipedes, insects and spiders	193
Egg masses	*197*
All washed up: natural marine debris	*200*
Glossary	*204*
Further reading	*210*
Index of common names	*212*
Index of scientific names	*215*

Preface and acknowledgements

This book is the fourth in a series of guides whose first edition was self-published in 1980 as a pocket-sized, black-and-white guide to rocky shores by authors Russell Synnot, Heather Powell and Geoff Wescott. The next two editions continued to cover only rocky shores, with a change in authorship each time (Gerry Quinn replaced Heather Powell and then Christine Porter replaced Russell Synnot) and the Victorian National Parks Association designing and editing the second and third editions. Over 25 000 copies have been sold.

The current authors thank the previous publishers and authors for their fantastic input, on which this edition draws and expands. This new edition elicits two new authors and a new publisher, and is significantly expanded to cover the main shore types in south-eastern Australia.

The authors acknowledge the Traditional Custodians of the Sea Countries that are covered in this book and the lands on which this book was written; pay our respects to their Elders – past, present and emerging; and thank them for taking such care in protecting and managing these seashores for tens of thousands of years.

The authors thank CSIRO Publishing for taking on this book and exposing it to a much wider audience than we have reached in the past. In particular, the authors thank Mark Hamilton and Tracey Kudis for their assistance throughout the writing of this book.

Most of the photographs in this guide are those of the authors, who are indebted to the following people for supplying images when their own collections were found wanting: Donna Squire, John Huisman, Jacqui Pocklington, Paul Carnell, Mel Wells, Peter Macreadie, Gerry Quinn, Rebecca Koss and Perry Davis. Donna Squire also converted the photographs in the plant section to the correct format for publication.

The authors are grateful to Rebecca Koss for contributing the section on citizen science and to John Huisman for taxonomic guidance for some of the seaweeds and seagrasses that appear in this guide.

Thank you to artist Masaru Matsuno for creating the drawings that improve the appearance of the key on page viii and to Xénia Keighley for the line drawings of *Durvillaea*.

We acknowledge the marine biologists, ecologists, taxonomists, students and the communities we have engaged with who have enriched our understanding of the marine environment and species diversity throughout our careers, thus indirectly contributing to this guide.

We would like to pay special tribute to our great mate, respected colleague and mentor, Professor Peter Fairweather (1958–2020). Peter was an influential Australian marine ecologist who made significant contributions to understanding the life on intertidal rocky shores. Some of Peter's earlier work involved research on the mulberry shell, *Tenguella (Morula) marginalba*, which is featured in this book. Peter had a broad ecological knowledge, extending well beyond the marine environment, which he used to inform environmental policy and management. He was an amazing mentor and inspirational teacher to many undergraduate and post graduate students. Ty Matthews and Alecia Bellgrove are honoured and eternally grateful to have had Peter as a mentor during their formative years as marine ecologists.

Start here to get the most out of this guide

This guide covers the more common species that are found in the intertidal zone on shores in south-eastern Australia: from Port Lincoln in South Australia to the Hawkesbury River in New South Wales and including all of Tasmania. The 'intertidal zone' refers to the seashore that is above water level at low tide and under water at high tide.

The content of this guide is organised to simplify the process of identifying and learning about the plants and animals that a shore rambler is likely to encounter. An important aim of the book is to help you do this with the minimum harm or damage to the plants and animals and the habitats in which you find them (see Conservation Code, p. 1).

The descriptions of the plants and animals you are likely to encounter are divided into three sections: plants that occur in the intertidal zone; animals typically found on rocky (hard) shores; and animals associated with sandy and muddy (soft) shores. Within each section, the species are organised according to the type (taxonomic group) of plant or animal.

Hard shores encompass rock platforms, boulder shores, rock rubble and stone flats, rock pools and human-made structures, such as pier pylons, breakwaters and seawalls. Read pages 59–62 to learn more about hard shore types. Soft shores encompass sandy beaches, mudflats, seagrass beds, saltmarsh and mangroves (including boardwalks through mangrove forests). Go to pages 158–164 to learn more about soft shores.

The simple guide on the next page, used in combination with the table of contents, will direct you to the relevant pages to search for the plant or animal you are attempting to identify. The beachcombing pages (pp. 200–203) include some additional plants, animals and objects that you may find washed up on the shoreline. Some intertidal animals lay many eggs close together on the shore. The egg masses you are most likely to find are covered on pages 197–199.

Field Guide to the Seashores of South-Eastern Australia

Key to seashore animals and plants

Use in conjunction with the 'Table of contents'. Remember that animals could be in the hard shore or soft shore section, depending which habitat you are visiting.

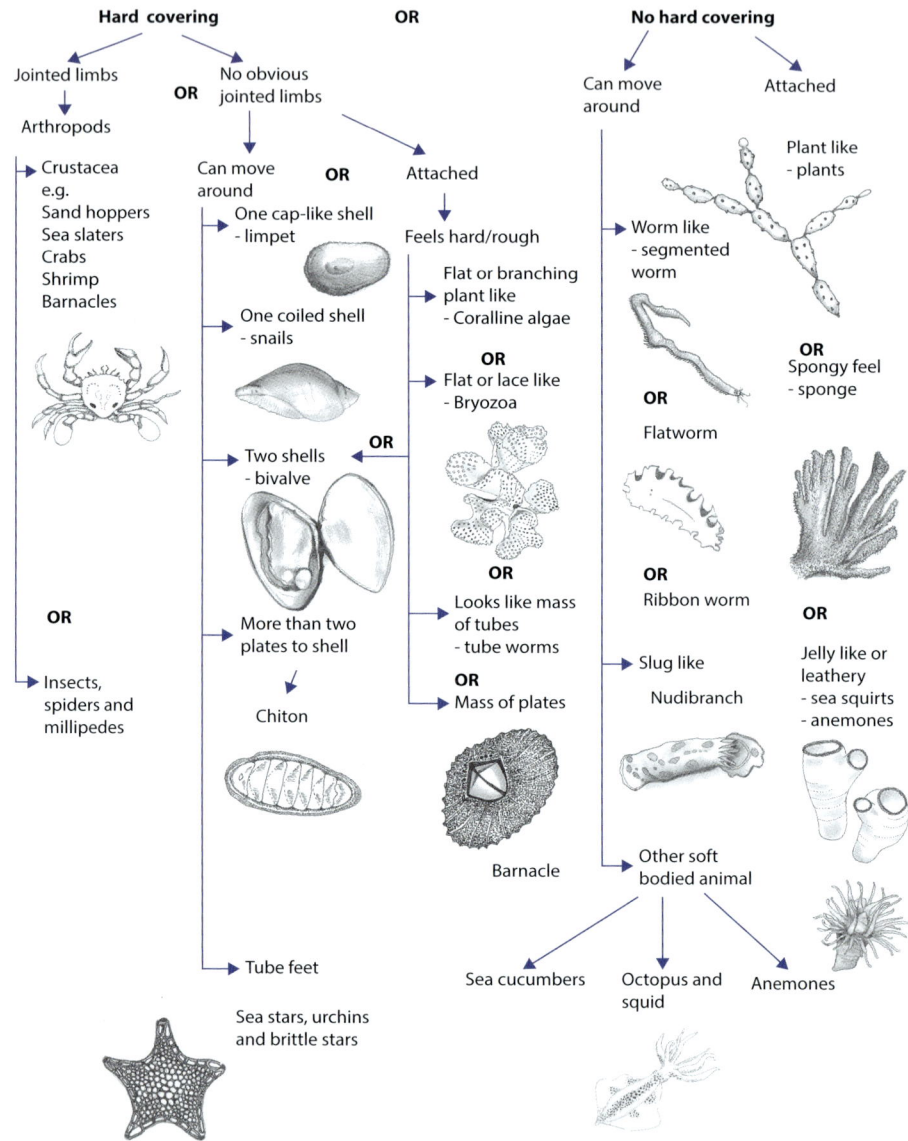

Illustrations by M. Matsuno.

In the interests of being accessible to a diverse readership, we have tried to minimise the use of technical language. Where we have used technical/scientific terms, you will find explanations for these terms in the glossary on page 204.

Finally, the map on each page provides an indicative species range within south-eastern Australia only. The Atlas of Living Australia provides more detailed distribution data for each species (https://www.ala.org.au).

Enjoy!

Introduction

Conservation Code

The plants and animals living on the edge of the sea are essentially marine in physiology and, despite their adaptations to inhabiting the land–sea interface, are very vulnerable to drying out (desiccation) and dying. Always think of this as you are enjoying your ramble on the seashore and follow our recommended Conservation Code.

Conservation Code

1. You are encouraged to observe and not handle the intertidal plants and animals you encounter.
2. Return everything you pick up or move, including stones and empty shells, to exactly where you found them.
3. Do not remove any living plant or animal from the shore.
4. Avoid trampling on plants or animals while walking across the shore.
5. Find out whether you are visiting a marine park, reserve or sanctuary, and make sure you are aware of the regulations governing these places.
6. Take your own rubbish away and pick up any rubbish you find, especially plastics and fishing line. These items can suffocate and strangle marine animals.

If you do need to look more closely at individual animals and plants, handle them with care where you find them (e.g. under the water surface in a rock pool). Remember to rinse sunscreen off your hands first. Do not remove whole plants (including algae) from where they are attached – they will rarely survive. Break off a small piece for closer observation, if necessary.

Note that shorelines within national parks are protected in the same way as the land component. If in doubt about the management status of a shore, assume it is a marine park and adhere to all the recommendations above. We encourage you to support community groups and government organisations that increase awareness of the marine environment and marine conservation.

Preparing for your visit to the shoreline

Seashores are fascinating places to explore at low tide. Although this activity is predominantly safe, a few simple steps will ensure you take care every time you visit.

Be mindful of two fundamental aspects of south-eastern Australia's seashores: tide and sea conditions.

- Know the tidal state before you go down to the shore and keep your eye on whether the tide is coming in or going out to avoid potential danger. Check the predicted low and high tide tables online using reliable weather apps, such as that of the Bureau of Meteorology (http://www.bom.gov.au/australia/tides/), or in daily newspapers or local shop windows before you set out.
- Sea conditions can change rapidly, along with the weather, and both can influence the tidal state. Every year in Australia people drown when they are washed off rock platforms or swept out to sea by waves and currents (rips). Check the weather forecast before heading out.

Be mindful of your safety when you head onto the shore.

- Be aware that parts of the shore can quickly be isolated by the incoming tide.
- Wear a hat, long-sleeved shirt and sunscreen to avoid sunburn.
- Wear sturdy footwear with ankle support and non-slip soles to reduce the chance of falls.
- Dress appropriately for the weather and conditions.
- Bend your knees and keep your head up when taking a closer look at animals and plants to avoid becoming lightheaded.
- Never place your bare hands or feet in pools, crevices or anywhere you cannot see into. Some animals living in these habitats have sharp spines that can inflict painful puncture wounds (e.g. sea urchins, p. 144), some can cause skin irritation (e.g. anemones, p. 64) and some are venomous (e.g. the blue-ringed octopus, p. 114, and cone shell, p. 107).
- Carry a mobile phone in case of emergencies.

If you are part of an organised excursion (school, university, field naturalist club etc.), abide by the organisation's safety protocols because your safety is the responsibility of the organisation running the visit.

To get the most out of your visit to the shore, you might like to pack a notebook and pencil, camera, magnifying lens and, of course, a copy of this book.

About scientific names

Most plants and animals have common names (e.g. limpet, dolphin) but a problem arises when common names for the same species vary across different states, Indigenous Nations, regions or countries. Scientific names are the same worldwide, regardless of variations in common names, and hence are more definitive in descriptions. In this book we always use the scientific name for each plant and animal, although we have included common English names as well.

The standard scientific way of naming an organism is by its genus followed by its species name (e.g. *Homo sapiens*; genus: *Homo*; species name: *sapiens*; common name: human). The genus and species names must be used together to name a species. Note that genus and species names are always written in italics. Individuals are grouped into species based on similarities in physical and genetic features and their potential, or actual, ability to interbreed. Species that have many features in common are grouped together in the same genus.

The species name represents the lowest taxonomic level within a taxonomic hierarchy, with the kingdom representing the highest taxonomic level followed by phylum. Each phylum is subdivided into class, order and family levels, before an individual is identified to genus and/or species level.

For each plant and animal featured in this book, we have provided at least one English common name followed by the genus and species name, phylum and class. For many we have also included order and/or family. We use 'spp.' to indicate that we are referring to more than one species from that genus and 'sp.' to refer to an organism from a particular genus whose species name is unknown.

The field of taxonomy (the classification of animals and plants into the various categories above) is a constantly evolving discipline: names and classifications can change as more is learnt about the evolutionary origins (e.g. through new fossil discoveries), physiology and genetics of particular species. We have used the currently accepted names according to the World Register of Marine Species (WoRMS; http://www.marinespecies.org), the Atlas of Living Australia (https://www.ala.org.au), Algaebase (https://www.algaebase.org) and the National Species List (https://biodiversity.org.au/nsl) at the time of publication. It is important to note that the taxonomic names that we have used in this book may have changed since publication, as relationships between species are studied further.

Species identification

Although photographs are the major aid to identification used in this guide, there are limitations to using obvious physical characteristics alone. It may not be possible to identify your specimen to lower taxonomic levels without inspecting some key characteristics under the microscope. We have specified in the text where caution is needed in this regard. Furthermore, colour pattern can be an unreliable trait on which to base identification. Individuals that belong to the same species may display variations in colour and external patterns (e.g. mottling, banding and striations). Animal examples include the sea star *Meridiastra calcar* (p. 148) and the marine snail *Cominella* spp. (p. 104). Similarly, macroalgae can change colour when exposed to direct sunlight. Red algae can be bleached and transition from red to orange or pink and ultimately to white, and brown algae exposed to ultraviolet (UV) light can release phenolic compounds (which act as natural sunscreens) that make them appear black.

Use this guide as the first step in the identification of the animal or plant you are looking at. If you would like a more precise identification after you have compared your specimen against the photograph and read the associated text, we suggest you then refer to the relevant authoritative text or scientific literature (some of which is listed in the Further Reading section).

Recognising carnivorous versus herbivorous snails: gravy boats and salad bowls

Many of the carnivorous marine snails have shell openings (apertures) with a characteristic shape that can be used to separate carnivorous and herbivorous snails. Carnivorous snails have a siphon through which they draw in water and pass it across a sensory organ that allows them to 'smell' and locate their food. This siphon emerges from a notch in the aperture, which leads to the shape of the aperture resembling that of a 'gravy boat'. If you find a snail that has an aperture that reminds you of a gravy boat, then you are likely to have found a carnivorous snail. These snails will either be predators or scavengers. In contrast to carnivorous snails, most herbivorous snails have round apertures, similar to a 'salad bowl'. We have illustrated this difference in the images below of *Cominella lineolata* ('gravy boat') and *Chlorodiloma odontis* ('salad bowl').

Apertures of the carnivorous snail *Cominella lineolata* (a) and the herbivorous snail *Chlorodiloma odontis* (b).

Sea Country and Saltwater Peoples

Australia is home to the longest continuous cultural history in the world (>65 000 years), with a richness and diversity of Aboriginal and Torres Strait Islander cultural groups (>250 known language groups). Indigenous Australian cultures are living cultures that are based on a rich body of knowledge, ways of thinking and doing and cultural practices. Indigenous Australians hold extensive traditional ecological knowledge and customary traditions surrounding the conservation and use of coastal resources, much of which has, unfortunately, been lost, fragmented or is 'sleeping'.

The geographical scope of this book encompasses at least 35 Aboriginal language groups whose Countries have a direct contemporary connection to the coast. Indigenous Australians from coastal areas across the nation, who are the Traditional Owners/Custodians of the Lands and Waters characterised by saltwater environments, are often referred to as Saltwater People. Saltwater Peoples have a Country-specific relationship to their particular Lands and Waters to which language is integral (e.g. the Gunditjmara Peoples will have different words and cultural associations to certain marine organisms to those of the Boon Wurrung Peoples some 250 km away, despite commonality of species). Thus, there is not one Aboriginal name and use for each species, but potentially many names and different uses for each species, especially where species ranges traverse many different Sea Countries (e.g. *Ecklonia radiata*, p. 47).

We recognise that Indigenous Australians have never surrendered any rights to their Land and Sea Countries, nor access to their cultural resources and traditional knowledge. We respectfully acknowledge that the coastal areas of south-eastern Australia were among the most densely populated regions of pre-colonial Australia, and that the Saltwater Peoples of this region share rich connections to, and knowledge of, these areas and species. The contemporary and historical uses and cultural significance of coastal resources (such as shellfish, marine mammals and finfish) to Aboriginal communities have been studied in several locations around Australia. This knowledge is not ours to share. It must be shared in collaboration with the Saltwater Peoples of the many Nations across this region and, indeed, is beyond the scope of this book. Hence, for common names, we use English names rather than selecting one of many Aboriginal names to use without consent.

As you explore the seashores of south-eastern Australia, you may well come across physical evidence of the importance of this region to Aboriginal Peoples, including coastal shell middens and sacred sites or artefacts. Check the online resources in our Further Reading section to learn more about how to recognise places that are culturally significant for Saltwater Peoples. In addition to conserving the environmental values of these lands and waters (see Conservation Code above), we ask that you respect the connections and cultural values for Aboriginal Saltwater Peoples and explore opportunities for all of us to conserve the historical and contemporary importance of Sea Country together.

Human impacts on seashores in south-eastern Australia
Climate change

The impact of human-induced climate change on our south-eastern shores is already dramatic and will likely become more pronounced over the next decade. Although a full discussion of these impacts is well beyond the scope of this book, an overview is given here with reference to further readings (p. 210).

Three major impacts stand out when discussing **climate change** on south-eastern shores. The most well-known impact is the rise in **sea level**. The sea is rising at an increasing rate. The main contributors to this rise are the thermal expansion of the oceans and increased melting of polar ice and glaciers. When combined with storm surges, offshore weather events and king (extreme) tides, significant coastal erosion is inevitable.

On soft-sediment, high-energy shores (sandy beaches usually) backed by dune systems, sea level rise will increase erosion of most shorelines. The beach may retreat higher but, if it is backed by private property or sea walls, for example, the extent of the 'beach' will be reduced or even eliminated.

Rock platforms are created as waves break and erode the base of cliffs and bluffs over long periods of time. This erosion process will not be fast enough to form new platforms if sea level rise proceeds at the predicted rates, and these platforms and their flora and fauna may disappear over time.

The second major impact of climate change is the increase in **sea temperature**. Oceans occupy approximately 70% of the surface area of the planet. Climate

Example of beach erosion from rising sea level.

change on land and in the atmosphere has been lessened because oceans have absorbed vast quantities of carbon dioxide and as much as 90% of heat trapped in the atmosphere. The average increase in sea surface temperature along the east coast of Australia has exceeded the global average for the past 50 years. This region is recognised as a hotspot for global warming.

As climate changes, species may move to stay within their preferred environmental conditions, resulting in a change to their distribution patterns, a process that has been called 'geographical range shift'. Marine plants and animals often disperse via planktonic stages in their life cycle, enabling them to move with ocean currents. One consequence of a rise in sea temperature has been a change in ocean currents. The East Australian Current flows the length of Australia's east coast, carrying warm water from the tropics into the Tasman Sea. This current has been moving further south over the past 50 years, resulting in southward changes to the distribution of some species. For example, the giant rock barnacle *Austromegabalanus nigrescens*, absent from Tasmania in the 1950s, is now recorded widely along Tasmania's east coast.

The impact of the extension of species ranges southward has been devastating in some areas. The decline in the extent of Tasmania's giant kelp forests has been attributed, in part, to expansion of the range and numbers of sea urchin species that eat kelp faster than it is able to regrow, creating 'urchin barrens'. Direct effects of temperate on kelps and the overfishing of urchin predators, such as crayfish, have also been implicated in the creation of urchin barrens.

The situation is more complex for intertidal and shallow subtidal flora and fauna living on south-facing coastlines, such as most of Victoria, southern Tasmania and some of South Australia. Here, migration further south to stay within an optimal water temperature range is simply not possible.

The third impact of climate change is considered by many ecologists as potentially the most severe: **ocean acidification**. As the sea dissolves more and more carbon dioxide, the pH (a measure of acidity versus alkalinity) is lowered, which, in layperson's terms, means the water becomes more acidic. Of particular concern is the impact of seawater acidification on the formation of calcium carbonate, which is the basis for seashells and many protective skeletal parts of a large range of algae and animals, including, for example, plankton (the microscopic plants and animals floating just below the surface waters) and benthic crustaceans. Plankton is the basis of most marine food webs, from microscopic zooplankton (animals) feeding on phytoplankton (algae) all the way through to the largest animal to ever live on Earth, the blue whale, which relies on krill and other small animals as its main food source. As increasing acidity hinders calcium carbonate formation, important plankton are likely to decline, leading to significant changes in marine food webs. This could be truly catastrophic, given the dependence of billions of people around the globe on fish and crustaceans as their primary food and protein sources.

Waste and pollution on our seashores

For many centuries, the sea has been viewed by humans as the ultimate waste dump: out of sight and out of mind. Some of this pollution comes from ships and boats, as a result of throwing rubbish overboard, leaking oil or flushing ballast water, but an estimated 70% of marine pollution comes from land-based sources (e.g. storm water and sewage effluent discharge).

Plastics pollution, including microplastics that are ingested by a range of marine organisms, has become a global problem. Plastics take decades to break down, so the impacts of this pollution will be long lasting, even after the elimination of single-use plastics. Plastics, together with other pollutants (from treated sewage effluent through to concentrated marine brine [a waste product of desalination plants], urban run-off and discarded fishing line) are widespread and pernicious. Human activity is the source of this waste and pollution, and hence it is the management of human use and land-based activity that is needed to reduce marine pollution. A simple way we can all help reduce marine pollution is to reduce litter – litter dropped in the street is washed down stormwater systems into creeks, streams and rivers, with most ending up in the sea.

Human impacts on the sea can result from practices that occur hundreds of kilometres inland. Examples include increased sedimentation from the clearing of native vegetation, pesticide pollution from agricultural practices and plastics pollution. We can all contribute to the health of our coasts and rivers, even if we do not reside beside the sea.

Other human impacts

Approximately 85% of the Australian population resides alongside estuaries or the open coast. Coastal development has major effects on shorelines, including through the potentially damaging effects of increased human recreation. People trample plants and animals on rocky shores and collect intertidal animals for food and fishing bait, resulting in significant population depletions of some species. Other human impacts on our shores include the compaction of sand from driving vehicles on beaches, major threats to nesting shore birds from dogs on beaches and interference with natural processes from engineering structures.

Conservation and management of south-eastern Australian shores

Responsibility for the planning, management and conservation of all intertidal shores in south-eastern Australia lies with the governments of South Australia, Victoria, Tasmania, New South Wales and the Australian Capital Territory (for a small section near Jervis Bay). In fact, the states have control of all land, freshwater and marine systems from the high tide mark to 3 nautical miles (~5 km) offshore under agreement with the Commonwealth (federal) government. The Commonwealth controls the sea area from 3 nautical miles offshore out to the edge of the Exclusive Economic Zone (EEZ) 200 nautical miles (~370 km) offshore. Australia's marine domain is twice the size of Australia's landmass and is one of the largest in the world.

State-based responsibility for intertidal shores means that laws and management approaches may be different for each jurisdiction – not something that is ecologically sound or easy to describe. Please check the relevant state and territory government websites, which can be found by searching for 'marine management', 'marine protected areas' (MPAs) and 'marine conservation and protection', for detailed and up-to-date information.

Marine parks (collectively called MPAs) have been established in south-eastern Australia, as elsewhere, as the cornerstone of conserving and protecting the shoreline environments and associated flora and fauna. According to the International Union for the Conservation of Nature (IUCN), marine parks are:

> *a clearly defined geographical space, recognised, dedicated and managed, through legal or other effective means, to achieve the long-term conservation of nature with associated ecosystem services and cultural values. (Fitzsimmons and Wescott 2016)*

The shores you visit while using this book may be within MPAs. It is your responsibility to find out whether you are in an MPA or not. Most MPAs have regulation and educational signage at access points to assist you.

Various types of MPAs have been established in Australia and around the world, ranging from those where there are few restrictions on permitted activities to areas that are highly protected against exploitation. The latter are often referred to as 'no-take' marine parks. These highly protected MPAs usually prohibit all forms of fishing as well as other extractive or environmentally damaging uses while allowing for non-consumptive public appreciation, visitation and enjoyment. In Victoria, for example, a series of 'no-take' or highly protected MPAs have been declared as Marine Sanctuaries and Marine National Parks. In global terms, these are Category 1 and 2 IUCN protected areas – the highest protection available. Small aquatic reserves and marine sanctuaries in South Australia, New South Wales and Tasmania also fit into these categories. In addition, across south-eastern Australia there are several categories of less-protected 'multiple-use' MPAs and fisheries reserves.

Globally, MPAs are the 'poor relations' of national parks and conservation reserves on land – every bit as important to conservation as land reserves, but far fewer in number and area. If you want to find further information on MPAs, the

CSIRO publication *Big, Bold and Blue* describes the MPAs in every state (see Further reading, p. 211).

The level of protection applying to the plants and animals on shores that are not in MPAs may vary between and within states, as do government agencies responsible for the planning and management of these shores. Fisheries authorities, park authorities and other public land management agencies, committees of management and local governments (councils) may all have a role. For example, Victorian Fisheries Authority regulations prohibit the collection of animals in Port Phillip Bay in water <2 m deep.

The resulting lack of clarity about management responsibility and regulations is one reason we prepared this book (and its predecessors): to aid conservation practices on all shore types. Please read the Conservation Code (p. 1) and protect these shores for future generations and the enjoyment of others now.

Citizen science
Rebecca Koss

Scientists and the general public have a long and successful history of user-generated content, well before arrival of the digital age. The public collects data and the scientific professionals provide the analysis. This is citizen science, defined as the voluntary collection and analysis of scientific data by the general public in collaboration with professional scientists. The term was coined in the early 1990s, but also goes by other names, such as community science, crowd science, volunteer monitoring and participatory research. The first modern-day citizen science project was started in North America by the ornithologist Wells Cook.

The key to any successful citizen science program is volunteer buy-in. But why would a citizen want to be involved? Some of the wonderful overarching aspects of voluntary participation in citizen science are listed below:

- Participating in citizen science provides a meaningful opportunity for any member of the public to learn about Science, Technology Engineering, and Mathematics (STEM) and, in turn, to improve their scientific literacy.
- Anyone can be a citizen scientist. You do not need a science degree to take part in a citizen science program. Indeed, you will be taught how to collect data using straightforward tools and methods.
- By taking part, volunteers have the opportunity to work alongside scientists and like-minded citizens. The data collected contributes to a greater collective knowledge set that advances scientific decision making and solutions, providing evidence that can underpin policies and legislation.
- Citizen science is a key building block for creating environmental stewardship, where volunteers feel a sense of connection to place, nature and community, knowing that their efforts are contributing to a greater good.

The Australian Citizen Science Association (https://citizenscience.org.au/) is a good starting point for anyone who wants to become a citizen scientist and is looking to join a project in their local area. Here's your chance to make an impact and be part of the solution. Citizen science allows all of us to be agents of change. Together, we can make a difference.

Intertidal ecology

The intertidal zone is the section of shore between extreme high tide and extreme low tide levels. The width of the intertidal zone varies with tidal range and the slope of the shore; gently sloping shores may have an intertidal zone measured in hundreds of metres, whereas the upper and lower limits of this zone may only be 1 or 2 m apart on steeply sloping shores. Intertidal environments support a diverse array of organisms that have adapted to the large changes in temperature, salinity and moisture resulting from regular cycles of inundation with salt water and exposure to air, as well as the erosive action of waves breaking on the shore.

The changing tides play a significant role in determining the distribution and abundance of marine flora and fauna in the intertidal zone. Shores of south-eastern Australia typically experience two tidal cycles of unequal heights daily with a tidal range of <2 m. The daily difference in height between low and high tides varies depending on the phases of the moon. The difference in tidal height is greatest for spring tides that occur during new and full moons, thus exposing the intertidal zone to air for the longest period during the tidal cycle, whereas the difference in tidal height is least for neap tides that occur during the first and third quarters of the moon. The tidal cycle shifts by 1 h each day, meaning that daytime low tides are progressively 1 h later each day. Although the daily tide

Wave-cut rock platform near Jan Juc beach, Victoria.

times are predictable, the actual tidal heights will vary depending on local conditions, such as wind and barometric pressure.

Waves are an important force shaping coastal landforms. The erosion of sedimentary rock to create shore platforms is an example of this. The force of waves breaking on shores also influences the types of plants and animals that live in the intertidal zone. Rocky headlands exposed to frequent large waves are inhabited by plants and animals that can secure themselves firmly to the rock (e.g. bull kelp, p. 40, and barnacles, pp. 116–122). In contrast, rocky shores in sheltered environments may be covered in fine sediments and associated deposit-feeding animals. Species diversity tends to be higher on shores where the exposure to waves is somewhere between these extremes (i.e. areas that experience intermediate levels of disturbance).

How do plants and animals find their way to the shore?

Most of the plants and invertebrate phyla living on the shore possess a microscopic planktonic stage that lives in the water currents for anything between minutes to months. Here, organisms will either drift or swim, and this can play an important role in dispersal, particularly for marine plants and animals that cannot disperse as adults because they are sessile or sedentary. These planktonic stages (e.g. spores, zygotes, seeds and larvae) may look very different from the adult form that they grow into after settling on the shore.

Some invertebrate groups have bypassed the larval stage altogether, producing 'crawl-away' juveniles that hatch directly from the egg and resemble the adult. The juvenile stages may develop in egg capsules laid on a rock or a plant, or they may be brooded by the parent. The snails *Cominella lineolata* (p. 104) and *Australaria australasia* (p. 106) are examples of the former, whereas the sea star *Parvulastra exigua* (p. 147) is an example of the latter.

How do intertidal plants and animals move around?

Intertidal plants and animals can be divided into three main groups based on modes of mobility, which, in turn, influence patterns of species distribution. The first group consists of sessile plants and animals, which typically have a mobile planktonic stage that settles permanently in one place. The main sessile groups are marine algae and plants, sponges, some molluscs (e.g. oysters, p. 113), serpulid worms (p. 74), barnacles (pp. 116–122), bryozoans (p. 143) and sea squirts (p. 155). The second group consists of sedentary invertebrates that appear sessile when exposed during low tide but are capable of relocating to other areas. Examples include sea anemones (pp. 64–68), chitons (pp. 79–81) and limpets (pp. 86–88). The third group consists of highly mobile invertebrates, such as crabs, amphipods, lobsters, shrimp and octopus. This group can actively seek out shelter as the tide recedes or to avoid predators.

Where do the species in this book fit in the food web?

All photosynthetic organisms contain chlorophyll a to harvest light and convert it to chemical energy that can be used in growth, reproduction and repair. The three broad groups of macroalgae (i.e. red, green and brown seaweeds) also possess different photosynthetic accessory pigments (responsible for their different colours) that allow these different groups to photosynthesise across varying light levels. Seagrasses, saltmarsh plants and mangroves – the marine angiosperms or flowering plants – contain similar pigments to the green algae from which they evolved. Collectively, these photosynthetic groups are known as the photoautotrophs, and they support the heterotrophs, namely herbivores (plant eaters), detritivores (detritus eaters) and carnivores (meat eaters). The heterotrophs can be further divided into different groups as listed below:

Grazers: scrape food from hard surfaces or plants using hard mouthparts (e.g. the toothed radula of gastropods and chitons, or the Aristotle's lantern of sea urchins)

Predators: directly eat other live animals
Marine invertebrates display some fascinating predator–prey interactions and diverse solutions to capturing and consuming prey. For example, sea anemones have tentacles jam-packed with stinging cells (cnidocytes) that paralyse their unsuspecting prey and ingest their food through a single orifice that serves as both mouth and anus. Some predatory snails drill a small hole through the shell of their prey (see photograph on p. 174). Some sea stars prize open bivalve shells (e.g. mussels) with their strong arms and sticky tube feet; they open the shells just enough to evert their stomach and secrete digestive enzymes to consume the now-defenceless bivalve.

Scavengers: consume dead or dying animals (carrion)
Scavengers are the marine equivalent to vultures and include some of the crustacean, polychaete worm, gastropod mollusc and sea star species. This feeding behaviour is exploited by recreational fishers, who drag a fish carcass along the sand along some parts of the east coast to attract beach worms (Onuphidae).

Detritivores and deposit feeders: typically feed on fine-particulate organic matter, which can include both decomposed plant and animal matter
Deposit feeders are abundant in soft-sediment habitats. Selective deposit feeders have specialised feeding organs, such as palps and tentacles (see spaghetti worm, p. 72) that stretch out along the sediment surface to pick up small organic particles. Non-selective deposit feeders ingest volumes of sediment and strip out the organic matter as it passes through their digestive tract. Faecal castings of arenicolid worms (p. 163) lying atop the sediment are the end product of a

non-selective deposit feeder, whereas the small round pellets deposited by soldier crabs and some ocypodid crabs are the end products of selective deposit feeders.

Filter/suspension feeders: ingest phytoplankton, zooplankton and small organic particles from the water
Examples of filter/suspension feeders include sponges, some polychaete worms, oysters and mussels, barnacles, feather stars and sea squirts. A diverse array of structures has evolved in these different organisms to capture and filter suspended food particles from the water.

Adaptations to life within the intertidal zone

Intertidal animals and plants require moisture for gas exchange (including respiration and photosynthesis), movement and feeding/nutrient uptake. Plants and animals living on rocky shores have adaptations that enable them to withstand short periods of exposure to air during the tidal cycle. The cell walls of seaweeds are composed of fibres (e.g. cellulose or xylan) embedded in a gelatinous polysaccharide matrix that can help the seaweeds survive desiccation by absorbing and holding water, and to withstand wave action by providing strength and flexibility. Mobile invertebrates can seek shelter in rock pools, crevices and under rock, whereas other invertebrates may rely on a hard protective shell and operculum, which also provide protection against predators.

On sandy and muddy shores, the moisture retained within the spaces between sediment particles and underneath wrack can buffer infaunal invertebrates against desiccation and high temperatures. In fact, some of the sandy beach invertebrates (insects, spiders and some crustaceans) will drown if submerged by water for too long, so they tend to be found at higher elevations on the shore.

Burrowing into the sediment affords many infaunal invertebrates with protection against predation, reducing the need for a heavy, robust shell. Some deep-burrowing, infaunal bivalves have thin, brittle shells, in contrast to the hard shells of their epifaunal relatives that live attached to hard surfaces. Similar observations are true for some of the burrowing crustaceans, such as the ghost shrimp. This is likely to allow these invertebrates to invest more energy in rapid growth and reproduction.

Distribution of marine plants and animals on the shore

Many intertidal species have marked upper and lower limits to their distributions in the intertidal zone, and an array of physical factors and biological processes interact to determine these limits. Physical factors include the duration of exposure to air, temperature and wave energy. Biological interactions include competition for space or food between individuals (of the same or different species), predation by other species and the settlement of planktonic propagules. Physical factors appear to be more important in determining the upper limits of species distributions, whereas biological interactions influence the lower limits. Features of shore topography, such as crevices, rock pools, patches of sediment and rock rubble, contribute to more complex distribution patterns.

Competition can be particularly important on rocky shores, where two-dimensional space is at a premium, but may not be as intense in soft-sediment habitats, where organisms occupy a three-dimensional habitat and can use different sediment depths to avoid direct competition for space. Another contrast between rocky intertidal and soft-sediment shores is that the continuously shifting sands and mud prevent the settlement of sessile biota, such as barnacles and rock oysters. Some sedentary animals can occur in soft-sediment habitats, where they are typically attached to a hard object, such as a clam or a rock lying just below the sediment. An example of this is the anemone *Anthopleura hermaphroditica* (p. 67).

On rocky shores, some distribution patterns related to height on the shore are common to temperate intertidal shores worldwide. The supralittoral fringe, or splash zone, is often inhabited by lichens (p. 58) and littorinid snails (p. 98). The upper intertidal zone is the highest section of shore regularly submerged during tidal cycles. The most obvious inhabitants of this zone are the air-breathing limpets (pp. 86–88) and barnacles (pp. 116–122). True limpets, grazing snails, oysters, mussels, tubeworms and barnacles are typical of the mid-intertidal zone. On platform shores this zone may be covered in large expanses of Neptune's necklace (p. 39). The lower intertidal zone is usually dominated by macroalgae and filter-feeding sea squirts (e.g. cunjevoi, p. 155). A diverse array of algae and marine invertebrate species live in the sublittoral fringe zone, exposed during very low spring tides, and in the near shore subtidal zone.

Distribution patterns on soft-sediment shores are generally not obvious to the casual observer: there are few plants, and most of the animals live within the sediments. Factors such as wave disturbance, particle size and moisture content contribute to where infauna are found on soft-sediment shores (see pp. 158–164).

Plants

Introduction to marine plants

The macroscopic plants of intertidal marine environments will be either seaweeds or flowering plants, but the size, shape and structure (the morphology) of these two groups differ. Flowering plants evolved on land with familiar plant structures including roots, stems, leaves and flowers, and specialised **vascular tissues** that transport water and nutrients from the soil throughout the plant. The leaves are the primary site of photosynthesis, a process by which carbon dioxide (entering through pores in the leaves from the air) and water (captured by the roots from the soil and transported to the leaves) are converted to sugars and oxygen in the presence of sunlight. The stems provide structural support and elevate the leaves towards the light. However, for marine flowering plants (seagrasses, saltmarsh plants and mangroves), there are challenges associated with their evolutionary return to the ocean, including dealing with waterlogged soils, high salinity, low carbon dioxide and low light relative to terrestrial environments.

In contrast, seaweeds evolved in the ocean, bathed in seawater, into three broad and diverse lineages, namely red (Rhodophyta), green (Chlorophyta) and brown (Phaeophyceae) algae. The colours of these algae are derived from various pigments used to capture different wavelengths of sunlight for photosynthesis. The body of a seaweed is called a **thallus** (plural **thalli**). Most seaweeds are attached to hard substrata (e.g. rocks, shells, pier pylons, other seaweeds) by an anchoring structure called a **holdfast**, but the holdfast does not take up nutrients like roots do (seaweeds take up nutrients directly from the water surrounding all their tissues). Holdfasts are usually constructed of either solid tissue, tangled fine **rhizoid** filaments or thick, cylindrical, intertwining branches called **haptera**. Strong natural glues attach the holdfasts to the substratum. Although all parts of a seaweed can capture light and photosynthesise, most photosynthesis occurs in the **blades**, which have high densities of chloroplasts (the structures in which photosynthesis takes place). A distinct flattened section on blades is known as a **lamina**. Some seaweeds have a structure analogous to a stem, called a **stipe**, that serves to elevate the blade towards the water surface and the sunlight. This is also aided by gas-filled **air bladders/vesicles** that act as floats in some species of brown algae. The stipe and blade together are called a **frond**. Fronds may be branched or unbranched, and the branching patterns are often important for identification, so require careful examination. The fronds in some species branch many times, and the small lateral branches are known as **ramuli**.

Seaweeds vary in growth form and include filamentous species (in which the thallus is formed from single or multiple fine filaments of cells in series), **coenocytic** species (in which the cross-walls are not formed during cell division such that the thallus consists of a huge, single cell with multiple nuclei), crustose forms (in which the thallus grows prostrate against the substratum) and arborescent forms (in which the thallus grows erect and branching). One group of green algae (*Caulerpa*) is coenocytic and grows a bit more like a grass, whereby, instead of a holdfast, it has a spreading **stolon** attached by small filaments called **rhizoids**. These stolons are superficially analogous to the spreading **rhizome** and roots of seagrasses (and some terrestrial grasses), although they are involved in attachment only and not in nutrient or water uptake.

Although marine flowering plants reproduce by the pollination of flowers, seaweeds have diverse and complex modes of reproduction, often involving alternating generations of **sporophytes** (which produce **spores** in **sporangia**) and **gametophytes** (which produce **gametes** in **gametangia**). Some species have gametophytes that are **monoecious** (hermaphroditic; with both male and female reproductive structures on a single thallus), whereas others are **dioecious** (with separate male and female thalli). There is a wide range of reproductive structures (both sporangia and gametangia) among the different algal lineages, and they are often important for identification, particularly in the red algae. This means it is sometimes impossible to accurately identify a specimen to species level unless it is fertile at the time of inspection. The gametangia that produce eggs are known as **oogonia**, whereas those that produce sperm are called **antheridia**. In some red and brown algae these gametangia are housed in an internal cavity with a small opening known as a **conceptacle**. Blade branches that are covered in conceptacles in some brown algae are called **receptacles**.

Blue–green algae

Rivularia firma
Phylum Cyanobacteria, Class Cyanophyceae, Order Nostocales

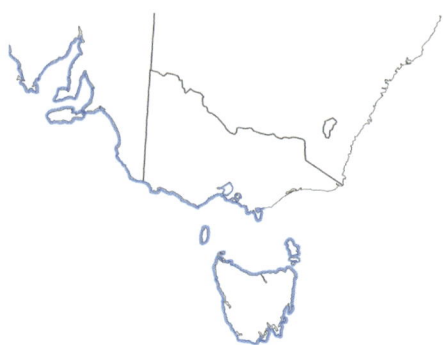

Rivularia firma.

Range
Two Peoples Bay, WA, to Wilsons Promontory, Vic., and Tas.

Appearance
Blue–green algae are among the very oldest organisms on Earth. They are actually types of photosynthetic bacteria that lack nuclei, called Cyanobacteria, and are not algae at all. Although many Cyanobacteria are unicellular, some species are able to form large colonies of cells that look like seaweeds. *R. firma* grows as firm but slimy, gelatinous, dark green to black globular colonies that are slippery when wet. The colonies comprise individual cells that are arranged into hair-like structures (known as trichomes) that are aligned vertically within in a jelly-like matrix and are easily distinguished under a microscope.

Habitat and ecology
This blue–green alga grows in the upper intertidal zone. It is seasonal, appearing from late spring through to autumn, but little is known of its ecology. Blue–green algae contain several chemicals that have potential commercial application, such as in pesticides, pharmaceuticals and sunscreens. *R. firma* is among the species that have been investigated for antibiotic, immunosuppressant, anticancer, antiviral, anti-inflammatory and photoprotective properties.

Red algae
Nori (Japan), laver (UK), karengo (New Zealand)
Porphyra/Pyropia spp.
Phylum Rhodophyta, Class Bangiophyceae, Order Bangiales, Family Bangiaceae

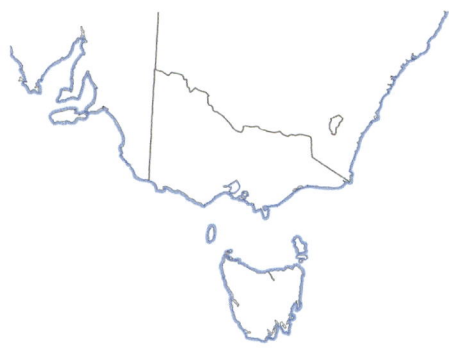

Porphyra sp.

Range
South-eastern Australia; globally in cool temperate regions

Appearance
Unlike all other red algae covered in this book (and most of the red algal taxa on our shores), nori belongs to the morphologically simple, and relatively small, class of Bangiophyceae. Superficially, the sporophytes of these seaweeds can look a bit like the green sea lettuce (*Ulva* spp., pp. 32–33), except that they are usually olive green to purple or reddish brown in colour. Nori have thin (only one cell thick), translucent, membranous blades that can be broadly ovate to elongated, sometimes tapering to a point or sometimes sickle shaped. The margins of different species can be smooth, toothed, wavy or ruffled. Distinguishing among species is not for the faint hearted.

Habitat and ecology
In southern Australia, nori can be found on intertidal rocky reefs in the colder months from late autumn/winter to early spring. Nori occurs in a microscopic filamentous phase over summer. This group of species is edible and highly valued in many parts of the world for its flavour and nutritional value – it can have up to 50% protein and a relatively high proportion of omega-3 eicosapentaenoic acid (EPA). Nori has been harvested and traded in Asia for thousands of years, cultivated in Japan since the 17th century, and global production is currently valued at around US$3 billion per annum. Yakinori sushi sheets are made by stacking and drying multiple thalli on top of each other in squares and then toasting them until crisp.

Coralline red algae
Phylum Rhodophyta, Class Florideophyceae, Order Corallinales

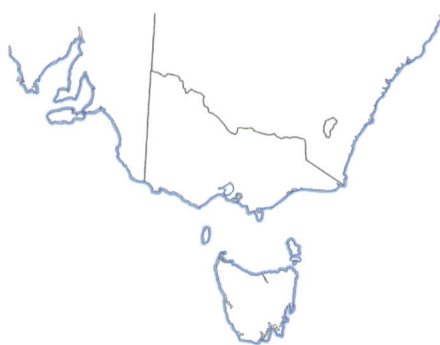

Range
Worldwide

Appearance
Some coralline algae have many-jointed (articulated) branching fronds, whereas others form encrustations (pink paint-like markings) on rocks. The cell walls of coralline algae are hardened by deposits of calcium carbonate, giving the plants a coral-like texture. Articulated coralline algae have non-calcified joints (called geniculae) between the calcified segments, allowing the thalli to move with water currents without snapping. Encrusting species usually grow prostrate over the substrata or as sheets (similar to some wood fungus or corals) and lack joints. Living plants are pink, purple or red, but those growing on rocky shores are often bleached white by the sun. Common genera of articulated corallines on southern Australian coasts include *Corallina*, *Amphiroa*, *Metagoniolithon*, *Cheilosporum*, *Haliptilon* and *Jania*, whereas encrusting forms are often represented by species from the *Lithothamnion* and *Porolithon* genera, among others.

The morphology of species can be greatly affected by the environmental conditions, often making reliable identification incredibly difficult. Confirmation by examination of the microscopic structure and reproductive structures, possibly in combination with molecular data, is usually required. However, there are a few species of articulated corallines common in south-eastern Australia that can be identified from photographs, at least to genus level, with careful examination of branching patterns.

Habitat and ecology
Both jointed and encrusting coralline algae are often common on intertidal rocky shores, in rock pools and subtidally to great depths. The calcification provides some protection against high UV exposure during low tide, as well as grazing by invertebrates. In deeper waters, these encrusting species can completely encase small rocks, pebbles or sand and become free-living, ball-like rhodoliths that can form extensive beds that become biodiversity hotspots. Like other calcifying organisms in the ocean, coralline algae may be threatened by increasing ocean acidification due to global climate change because the integrity of the calcified walls may be compromised.

Corallina officinalis
Family Corallinaceae

Range
Temperate regions worldwide

Corallina officinalis can form biodiverse, densely intertwined turfs (height 10–50 mm) in the intertidal zone to withstand strong wave action and desiccation but, subtidally, the same species can be arborescent (commonly to a height of 60 mm or more). *C. officinalis* has a crustose holdfast from which multiple fronds arise. The fronds have a cylindrical central axis and repeated pinnate (feather-like) branching. At branching points, the segments are more or less trapezoid with a broader top than base. When fertile, the reproductive structures appear in conceptacles at the tips of lateral branches. *C. officinalis* has been used medicinally as an antacid and as a bone-forming material, as well as in cosmetics.

Amphiroa spp.
Family Lithophyllaceae

Range
Worldwide in tropical, subtropical and temperate regions

Amphiroa species are common in rock pools and the subtidal zone across southern Australia.

Branching occurs in a dichotomous (forked) pattern and segments are usually characteristically flattened, with blunt to slightly rounded tips that are often bleached white, particularly in shallower waters. When fertile, the reproductive structures are unusually randomly scattered on the segments.

Metagoniolithon stelliferum
Family Porolithaceae

Range
Abrolhos Islands, WA, to NSW and Tas.

Metagoniolithon, endemic to southern Australia, is characterised by dichotomously branched cylindrical axes that have a series of branches arising from around the joints. *M. stelliferum* is common in rock pools and the shallow subtidal zone on moderately wave-exposed rocky coasts. It often grows epiphytically on seagrasses and, less commonly, on large brown algae.

(a) *Corallina officinalis* (subtidal arborescent form), (b) *Amphiroa* sp., (c) *Metagoniolithon stelliferum* and (d) encrusting coralline algae (photographs b, c, d by D. Squire).

Turfing red alga
Capreolia implexa
Phylum Rhodophyta, Class Florideophyceae, Order Gelidiales, Family Gelidiaceae

Capreolia implexa (photographs by D. Squire).

Range
Wittelbee Point, SA, to Broken Bay, NSW, Tas., New Zealand, Chile

Appearance
This species grows as prostrate, entangled turfs, forming small clumps to extensive, irregularly shaped patches up to 30 mm thick. It can be golden brown to reddish brown in colour. The blades are cylindrical to flattened (<5 mm wide) with irregular branching, usually tapering at the tips. The holdfast consists of massed rhizoids on the underside of branches.

Most previous records of *Gelidium pusillum* occurring as intertidal turfs in southern Australia are most likely misidentified *C. implexa*. The life history of *C. implexa* is unique among the Gelidiales in having only two life phases (rather than the usual three life phases characteristic of most red algae).

Habitat and ecology
Common in the mid- to low intertidal zone on moderately to highly wave-exposed coasts.

C. implexa grows on exposed rock, amid *Hormosira banksii* mats (p. 39) and *Corallina officinalis* turfs (p. 20), and on mussels and *Galeolaria caespitosa* (p. 74). It can also grow on the shells of mobile molluscs, such as limpets (pp. 86–88), on rocky reefs and on the aerial roots and trunks of the grey mangrove *Avicennia marina* (p. 57), in wave-sheltered habitats. The turf structure withstands wave action and resists desiccation at low tide by holding water within the turf, and can provide important habitat for small, mobile invertebrates.

Banded goblets
Ceramium flaccidum
Phylum Rhodophyta, Class Florideophyceae, Order Ceramiales, Family Ceramiaceae

Range
WA, SA, Vic., NSW, Tas.

Appearance
This species forms very fine (<3 mm in diameter) filamentous tufts (up to 10 cm high) that are light red to dark reddish brown. When viewed with a hand lens, they have a distinctively striped appearance, with clear bands alternating with red bands, and goblet-shaped forked tips.

Habitat and ecology
This species is a common epiphyte on many different algae (especially on *Haliptilon*), but also grows on pebbles and rocky reefs. There are many other species of *Ceramium* that occur as epiphytes with which *C. flaccidum* could be confused. However, *C. flaccidum* is most conspicuous in the low intertidal zone on rocky reefs with elevated nutrient levels, such as those close to sewage wastewater discharge points, where it can be extremely abundant, occurring as small (commonly to ~50 mm high) filamentous tufts associated with turfs of *Capreolia implexa* (p. 22) or *Corallina officinalis* (p. 20).

Ceramium flaccidum (photographs by D. Squire).

Red feather weed
Ballia callitricha
Phylum Rhodophyta, Class Florideophyceae, Order Balliales, Family Balliaceae

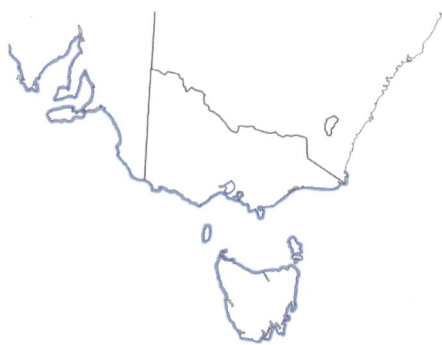

Range
Nuyts Reef, SA, to Green Cape, NSW, Tas.; isolated record from Geographe Bay, WA; New Zealand, South America, South Africa and Subantarctic Islands

Appearance
This species has a feathery habit with a conical holdfast (2–15 mm wide) formed from entwined rhizoids. It is generally medium to dark reddish brown but can become more orange to yellow (particularly the tips) if bleached by the sun when washed up on the beach. Thalli are highly branched and can grow to 360 mm in height. The fronds are flattened in a single plane, and pinnate (feather-like) branching occurs with opposite pairs on either side of a central axis. Multiple orders of branching can occur, but longer (5- to 10-mm) lateral branches occur irregularly along the central axes. Final ramuli are usually short (<1 mm), and cells on the central axes are strongly banded, giving a fish spine effect, especially if viewed under a hand lens or microscope.

Habitat and ecology
Very common on relatively deep rocky reefs or in the understorey of canopy-forming brown algae,

Ballia callitricha (photographs by D. Squire).

and sometimes in deep rock pools across southern Australia. *B. callitricha* is often washed up in shallow rock pools or on the beach. Thalli will often be seen with other red algae, including crustose coralline algae, growing epiphytically on the thalli of *B. callitricha*.

Green algae

Caulerpa
Caulerpa spp.
Phylum Chlorophyta, Class Ulvophyceae, Order Bryopsidales

Range
Tropical to subtropical globally, Mediterranean region and temperate Australia

Appearance
Caulerpa species grow erect fronds from a horizontal creeping stolon that spreads over the substratum with rhizoids for attachment. There are 104 currently accepted species of *Caulerpa*, with the highest diversity (33 species) in the world occurring in southern Australia. Species are distinguished by branching pattern and the size and shape of the smallest branchlets (known as ramuli), and are often distinctive and easily recognised from photographs. *Caulerpa* species are coenocytic.

Habitat and ecology
Caulerpa can grow attached to hard substrata like most other seaweeds, but are a bit unusual in that their creeping stolon allows them to also grow in the unconsolidated sand of seagrass meadows or wave-sheltered areas. They can grow from the intertidal zone down to a depth of at least 50 m, with species-specific distributions. *Caulerpa* species can be grazed by fish and sea slugs (nudibranchs) but have incredible healing powers and can heal wounds in minutes. Some species produce the hazardous toxin caulerpicin to deter grazers, but others, like *C. lentillifera* (commonly known as green caviar or sea grapes), are safe to eat and delicious.

Examples of common *Caulerpa* species in south-eastern Australia
Caulerpa brownii

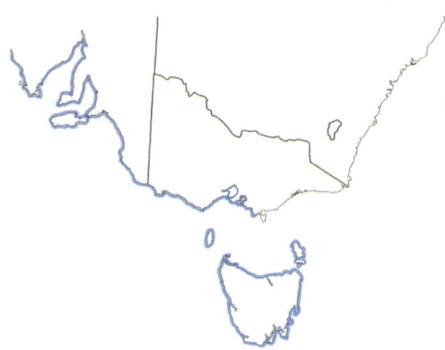

Range
From WA to Walkerville, Vic., and around Tas.; New Zealand, including Rakiura/Stewart, Tini Heke/Snares and the Rēkohu/Wharekauri/Chatham islands

Caulerpa brownii (photograph by D. Squire).

C. brownii is reasonably distinctive, with medium to dark green, (generally) unbranched, cylindrical fronds that are between 3 and 40 cm long and between 3 and 8 mm in diameter. The fronds are circled by dense, short (0.5–2.5 mm) ramuli that are bifurcated (forked twice) at the base (although this can be hard to see). *C. brownii* is common on exposed reefs from the subtidal fringe zone to a depth of approximately 40 m, as well as in low-shore rock pools, and can form dense clumps with intertwined stolons.

Caulerpa flexilis

Range
Geraldton, WA, to the Northern Beaches, NSW, and around Tas. and New Zealand

The fronds and ramuli of *C. flexilis* are similar to those of *C. brownii*, but in this case the fronds are branched with opposite second-order laterals in two dimensions. The ramuli encircle the central axis and lateral branches, are forked near their base and are 1–3 mm long. The fronds are usually 5–40 cm long. *C. flexilis* is common in rock pools and the subtidal zone (to a depth of 40 m) on reefs with moderate to high wave exposure.

Caulerpa flexilis (photograph by D. Squire).

Caulerpa longifolia

Range
Eucla, WA, near the SA border to Waratah Bay, Vic., and around Tas.

As the name suggests, the fronds of *C. longifolia* are encircled with long (5–15 mm) ramuli that are usually dark green and arise from the central axis in five rows (occasionally four or six rows), cylindrical, tapering and curved upwards. The erect fronds are also long (15–65 cm) and rarely branched. Found in rock pools and subtidally to a depth of 40 m on coasts with high wave exposure.

Caulerpa longifolia (photograph by J. Pocklington).

Caulerpa trifaria

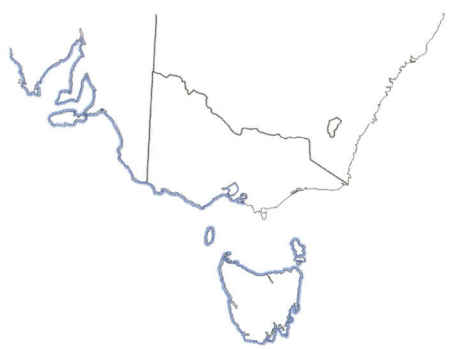

Caulerpa trifaria (photograph by J. Huisman).

Range
Cottlesloe, WA, to Western Port Bay, Vic., and around Tas.

The central axis of the erect fronds is often a bright yellowish green and the ramuli are a medium green. The fronds are usually unbranched, 5–25 cm long and 4–12 mm across. The ramuli are cylindrical and tapering, 3–9 mm long and arise from the central axis in three rows (like a Mercedes-Benz symbol in cross-section). This species can be confused with *C. longifolia*, which can occasionally have four (commonly five) rows of ramuli rather than three. It can be found on jetty pylons, rocks and sand in rock pools and subtidally (at depths of 2–31 m) on moderately wave-sheltered to wave-protected shores.

Caulerpa cactoides

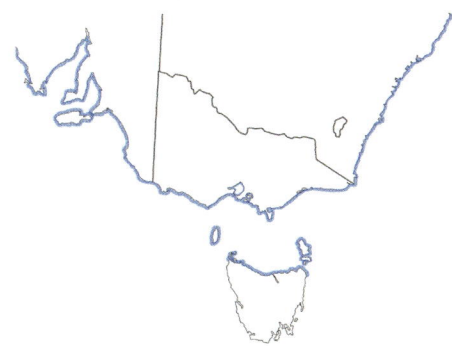

Caulerpa cactoides (photograph by J. Pocklington).

Range
Geraldton, WA, and SA, Vic., NSW and northern Tas.

The species name means 'cactus like'. *C. cactoides* grows as erect, emerald-green fronds with large oval to pear-shaped vesicles ('bubbles') as branches. There are also distinctive rings at the base of the upright fronds. This species is found on sheltered to moderately wave-exposed shores in the subtidal fringe zone and in rock pools.

Caulerpa taxifolia

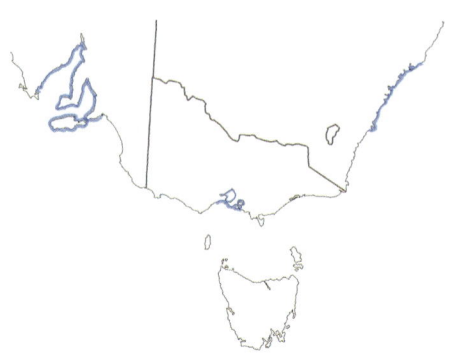

Caulerpa taxifolia (photograph by J. Huisman).

Range
Native to tropical and subtropical waters; invasive in 11 southern Australian estuaries

Caulerpa taxifolia, a fast-growing species native to tropical and subtropical waters, has been introduced via the aquarium trade and become invasive in temperate regions of Australia, the Mediterranean and California, impacting on native species and ecosystem function. The ability of this species to spread rapidly (including by fragmentation) in temperate waters has led to the listing of *C. taxifolia* as a noxious species in southern Australia.

Sea apples and green sea velvet
Codium spp.
Phylum Chlorophyta, Class Ulvophyceae, Order Bryopsidales

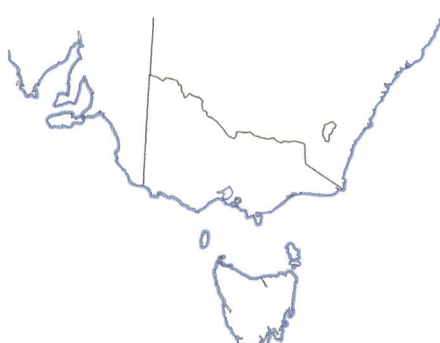

Range
Temperate and subtropical waters

Appearance
Codium species are coenocytic. There are two broad growth forms: in one, the thallus is flattened on the substratum or almost spherical (sea apples) with a large rhizoidal holdfast area; alternatively, the thallus is erect and usually branched, with a small holdfast that forms a spongy basal disk (green sea velvet). The thallus construction of both forms is a bit like a spongy pompom. The chloroplasts are concentrated within the swollen tips of each branch (called utricles) and there are so many chloroplasts full of chlorophyll that the thalli appear a very dark green to almost black. The appressed tips of the utricles covering the surface give *Codium* species a velvety appearance.

There are currently 144 accepted species of *Codium*, with the highest diversity occurring in countries like Australia (18 species) that span temperate and subtropical zones. Beyond separating the species into flattened/spherical versus erect and branching forms, identification to species level usually involves microscopic examination of the utricles and information about the distribution. For example, two species of sea apples that are common on Australian southern shores are *C. mamillosum* and *C. pomoides*. Both are dark green, firm, nearly spherical and attached to rocks by a mass of rhizoids. However, *C. mamillosum* is usually a bit smaller than *C. pomoides* (up to 8 *vs* 12 cm in diameter, respectively) and has larger utricles than *C. pomoides* (0.45–1 *vs* 0.09–0.125 mm in diameter, respectively). Similarly, there is high diversity and range overlap of erect branching species, for which reliable identification to species level usually comes down to utricle size, shape and the presence or absence of a cap (called a mucron), thus requiring microscopic examination.

Habitat and ecology
Codium spp. grow on intertidal and subtidal reefs down to depths of at least 40 m, where their high chlorophyll concentration, accessory pigments and chloroplast density enable them to harvest the small amount of light available at that depth. Flattened and spherical species generally occur in the low shore and subtidal fringe zones, in rock pools and under shaded rock surfaces, on shores of moderate to high wave exposure. Erect and branching species generally occur in the low-shore zone and below, as well as in rock pools, and are more common on coasts with moderate to low wave exposure. Although *C. fragile* is native to Australia, the weedy subspecies *C. fragile* subspecies *tomentosoides* is native to Japan, but has been introduced to the North Atlantic, northern Europe, Mediterranean and South Pacific, including Australia. This subspecies has vigorous growth and asexual reproduction and has become highly invasive and problematic for aquaculture, where it fouls nets and smothers oysters and other shellfish.

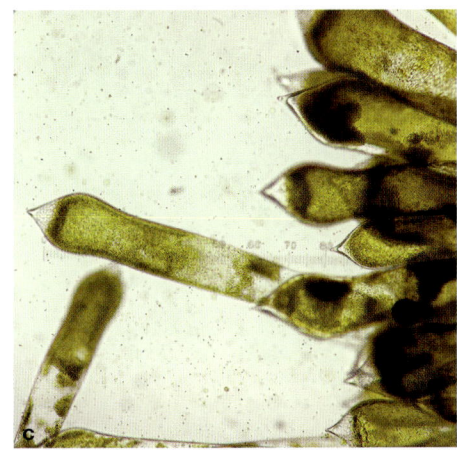

(a) *Codium* sp., (b) *Codium pomoides* and (c) utricles of *Codium* (photograph a by D. Squire).

Mermaid's necklace
Chaetomorpha coliformis
Phylum Chlorophyta, Class Ulvophyceae, Order Cladophorales, Family Cladophoraceae

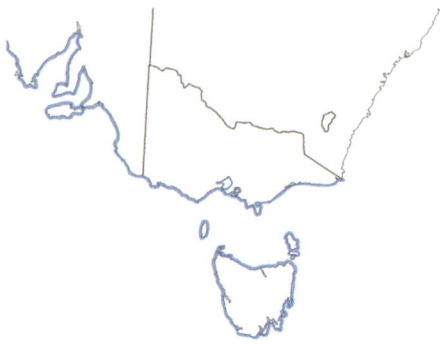

Range
SA, Vic., Tas.

Appearance
Chaetomorpha species, like other species in the Cladophorales, are constructed of filaments of single cells in series. Although some related genera have branching filaments, *Chaetomorpha* species are unbranched. The thallus of *C. coliformis* looks like a long (to 60 cm) shiny, green beaded necklace arising from a small holdfast and elongate basal cell, where each 'bead' is a very large cell (to 5 mm in diameter) that is easily visible to the naked eye.

Habitat and ecology
C. coliformis is often epiphytic on seagrasses (pp. 49–52), *Halopteris* (p. 37), *Ballia* (p. 24) and other algae, and occurs attached to the rocky reef from the very low shore to a depth of approximately 4 m, as well as in rock pools. It is relatively common on reefs with moderate to high wave exposure. *C. coliformis* is edible, with a taste reminiscent of cucumber and a surprisingly crunchy texture given its delicate appearance.

Chaetomorpha coliformis (photographs by D. Squire).

South African cladophora
Cladophora prolifera
Phylum Chlorophyta, Class Ulvophyceae, Order Chladophorales, Family Cladophoraceae

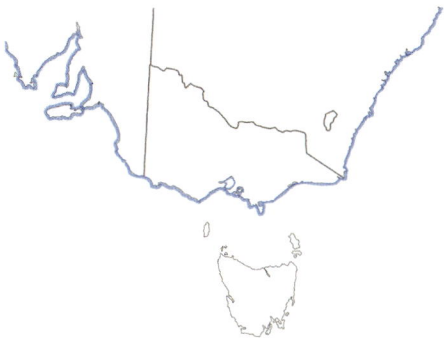

Cladophora prolifera (photograph a by J. Pocklington).

Range
Patchy distribution across WA, SA, Vic. and NSW; Gold Coast to Hopevale, Qld

Appearance
The plant body is dark green and 2–13 cm high. *C. prolifera* grows as stiff, fairly erect, tufts of densely branched filaments of single cells in series that have a coarse texture. It differs from other species of this genus, which tend to be softer and a lighter colour.

Habitat and ecology
C. prolifera is an introduced species native to South Africa that is now abundant in the mid- to low intertidal zone on many wave-exposed shores in southern Australia. This species has previously been incorrectly recorded as *C. rugulosa* in Australia.

Liverwort seaweed
Dictyosphaeria sericea
Phylum Chlorophyta, Class Ulvophyceae, Order Cladophorales, Family Siphonocladaceae

Dictyosphaeria sericea.

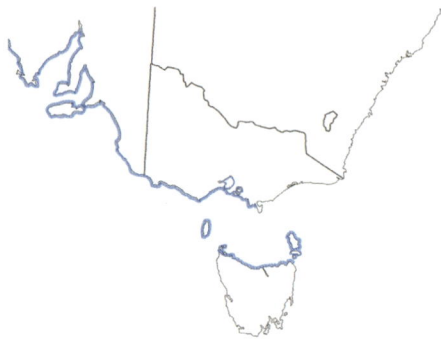

Range
Rottnest, WA, south to Walkerville, Vic., and northern Tas.

Appearance
The thallus (body) of this plant is an aggregation of large polygonal cells that form irregular, wavy, firm lobes attached to the substratum with rhizoids (root-like structures), giving it an overall appearance similar to a liverwort.

Habitat and ecology
This common species grows at low tide in shady areas, such as on the underside of overhanging rocks. Thirteen species are currently recognised for this genus, but *D. sericea* is the only one found in southern Australia.

Sea lettuce and green bait weed
Ulva spp.
Phylum Chlorophyta, Class Ulvophyceae, Order Ulvales, Family Ulvaceae

Range
Worldwide

Appearance
The fronds of *Ulva* are either broad and flattened (sea lettuces) or tubular and flattened (e.g. green bait weed). Some have slightly expanded margins, giving a ruffled appearance, and some are perforated. Plants are anchored to the substratum by a small, rhizoidal holdfast. Size varies among species, with some growing up to 30 cm long. There are currently 86 accepted species of *Ulva*, with 14 in south-eastern Australia. In most cases, microscopic examination is required to distinguish species. The taxonomy of this genus has been controversial, but it is now generally accepted that the broad-bladed, lettuce-like species and the more tubular or elongate species (which were previously classified in the genus *Enteromorpha*) both sit within the genus *Ulva*. These two growth forms are indistinguishable as juveniles: they all begin as a hollow tube with a single layer of cells (think of a tubular balloon). At some point during growth, the tube collapses in on itself (like a deflated balloon) so that there is now a flat blade consisting of two layers of appressed cells; all cells have a large, single

cup- or C-shaped chloroplast that can be seen under the microscope. Because there are only two cell layers, the blades are translucent, and usually a vibrant green (from the chlorophyll in the chloroplasts).

Ulva species cycle between two phases (gametophytes and sporophytes) that are morphologically similar, with all cells in the blades capable of producing either gametes (in the gametophyte phase) or spores (in the sporophyte phase). The marginal cells usually become reproductive and, when mature, the flagellated gametes or spores burst out of these cells, leaving behind a white outer margin. This can often be seen in the field, and sometimes lucky shore-goers may witness spawning events in small rock pools on warm days, with bright green swarmers seen swimming in the water and white margins on the blades of parent plants.

Habitat and ecology

Ulva species grow attached to hard substrata. On rocky shores, they are generally found in the low intertidal and subtidal zones and rock pools because they have little resistance to drying out; in wet areas they can be found high on rock platforms, in areas where freshwater seeps over intertidal rocks. They can also be found growing on structures such as pier pilings, boat moorings and hulls, and epiphytically on seagrasses and hard-shelled invertebrates. *Ulva* species are common in disturbed environments and areas with anthropogenic pollution and are tolerant to fluctuations in salinity. In areas where nitrogen levels are high (e.g. near river mouths or sewage effluent outfalls), they are able to produce a lot of chlorophyll, giving them a darker, but still translucent, green appearance. Conversely, if nitrogen is in low supply, they can be more yellowish. The fronds are eaten by some fish and crustaceans, whereas young plants are eaten by herbivorous snails. *U. compressa*, known as green baitweed, is used as bait to catch blackfish/luderick (*Girella tricuspidata*). *Ulva* species have few structural defences against grazing animals, but they can grow rapidly and reproduce prolifically to persist despite grazing pressure. They recruit in autumn and winter, at which time they are usually microscopic, with macroscopic thalli more abundant in spring and early summer. In late summer after reproducing (or after extreme low tides that coincide with hot weather/high UV), the blades will bleach completely and then die off. Although flavour differs among species, all *Ulva* species are edible, either fresh or dried and flaked (e.g. aonori in Japanese cuisine). However, care should be taken to forage away from urban areas,

Two *Ulva* species (photographs by D. Squire).

river mouths and other areas of potential pollution sources, where *Ulva* species can often be lush and abundant, but accumulate toxins. *Ulva* species are also being grown in wastewaters from prawn farms in Queensland to strip the nutrients (and any potential toxins) before effluent is discharged to the ocean, helping protect the Great Barrier Reef.

Brown algae
Encrusting brown algae
Phylum Ochrophyta, Class Phaeophyceae

Appearance
These algae appear as dark brown to black crusts with a fairly smooth texture and can be very slippery when wet. On southern Australian shores these algae are most commonly represented by *Pseudoralfsia verrucosa* and *Scytosiphon lomentaria*, but the two can be difficult to distinguish without microscopic examination, even though they belong to different taxonomic orders.

Habitat and ecology
Encrusting brown algae are common on rocks throughout the intertidal region. They are fairly resistant to grazing and trampling, so can be more abundant on shores with high foot traffic. Both *P. verrucosa* and *S. lomentaria* are common globally in temperate and colder waters, with overlapping distributions in southern Australia.

Limpet paint
Pseudoralfsia verrucosa
Order Ralfsiales, Family Pseudoralfsiaceae

Range
Cowaramup Bay, WA, to Broken Bay, NSW, and Tas.

Pseudoralfsia verrucosa forms medium to dark brown, round to irregular crusts that can grow to 5 mm thick and 50 mm across, often with slightly concentric bands. *P. verrucosa* attaches tightly to the rocks in the intertidal zone and often grows on mollusc shells (e.g. on limpets). Previously classified as *Ralfsia verrucosa*, *P. verrucosa* was recently redescribed with the aid of molecular and morphological data.

Chipolata weed, leather tube
Scytosiphon lomentaria
Order Ectocarpales, Family Scytosiphonaceae

Range
Cottlesloe, WA, to Sydney, NSW, and Tas.

Scytosiphon lomentaria alternates between two life cycle stages with very different body forms: *Ralfsia*-like encrustations (2–50 mm across) and unbranched tubular, medium to dark brown fronds (to 76 cm long, usually 1–5 mm in diameter) that appear on moister areas of the shore during winter.

(a) Encrusting brown alga (bordered by *Hormosira banksii*) and (b) *Scytosiphon lomentaria*.

Globe algae
Colpomenia spp.
Phylum Ochrophyta, Class Phaeophyceae, Order Ectocarpales, Family Scytosiphonaceae

Colpomenia (photograph by D. Squire).

Range
Worldwide

Appearance
The sporophytes of *Colpomenia* species grow as globular, balloon-like membranous thalli that can become convoluted and irregular in shape, usually not more than twice as high as they are wide. They are most commonly golden brown to yellowish in colour and can appear semitransparent. This macroscopic sporophyte phase alternates with a microscopic gametophyte phase or can develop without fertilisation (parthenogenesis).

The taxonomy of this genus is controversial, and it is often difficult to distinguish species without microscopic examination of reproductive structures and cellular structure, or DNA sequence information. There are currently 10 species recognised, with four of those occurring in south-eastern Australia: *C. sinuosa*, *C. peregrina*, *C. ecuticulata* and *C. claytoniae*. Although *Leathesia difformis* has a similar appearance to *Colpomenia* species, it has a slimy texture and is more gelatinous.

Habitat and ecology
Sporophytes of *Colpomenia* species are often common in the mid- to low intertidal zone of rocky shores (particularly in summer) or growing epiphytically on seagrasses or other algae. The bladder-like masses are attached to rocks or other plants by an irregular crusty base and retain seawater within the cavity, which helps reduce desiccation during low tide. *C. sinuosa* has high levels of natural antioxidants with potential in the food and pharmacological industries. *C. peregrina* was most likely introduced to Australia and New Zealand from the Asia–Pacific via fouling of fishing vessel hulls, with the first report of *C. peregrina* in Australia in 1967.

Dead-man's fingers, sausage weed
Splachnidium rugosum
Phylum Ochrophyta, Class Phaeophyceae, Order Scytothamnales, Family Splachnidiaceae

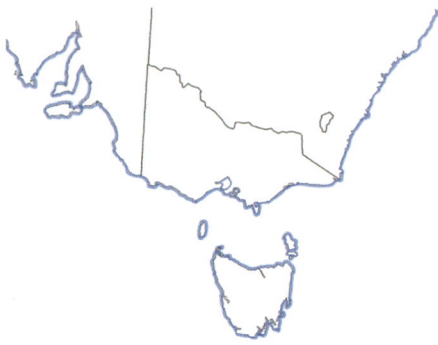

Splachnidium rugosum (photograph by G. Quinn).

Range
SA, Vic., NSW, Tas.

Appearance
Cylindrical (finger-like) wrinkled fronds up to 20 cm long (likened also to a clump of sausages) radiate from a holdfast. Some small lateral branches are present, reminiscent of a caveman's club. The plant has a jelly-like feel because it is filled with gelatinous material (rich in fucoidan, a sulfated polysaccharide with a fucose backbone that is embedded in the cell walls). This is the only species in this genus and easy to identify from photographs.

Habitat and ecology
Splachnidium is found in the mid-intertidal zone on exposed shores in cold temperate regions of the Southern Hemisphere. Visible sporophytes of this species appear in late spring, are most abundant through summer and overwinter as microscopic threads among barnacles and tubeworms. The fucoidan chemicals extracted from this species have been shown to have potency against the herpes simplex virus that is associated with the development and progression of Alzheimer's disease. Fucoidan extracts from *S. rugosum* are being further investigated as a potential treatment to slow or stop the progression of Alzheimer's disease.

Cousin Itt weed
Halopteris paniculata
Phylum Ochrophyta, Class Phaeophyceae, Order Sphaceriales, Family Stypocaulaceae

Range
Kangaroo Island and Port Willunga, SA, to Newcastle, NSW, and Tas., New Zealand, Chile, Subantarctic Islands

Appearance
Dark brown, densely tufted fronds (commonly 10–20 cm long) arise from a rhizoidal, often tangled, holdfast, giving a hairy appearance reminiscent of Cousin Itt from the Addams Family television series. Highly branched and slender (<1 mm in diameter) laterals are translucent with a dark spot on the end of paler tips. Previously recorded as *Halopteris gracilescens*, among other synonyms.

Habitat and ecology
H. paniculata is common in the lower intertidal zone, rock pools and down to depths of 13 m on rocky reefs in south-eastern Australia.

Halopteris paniculata (photographs by D. Squire).

Brown fan weeds

Padina fraseri, *Zonaria* spp., *Lobophora* spp.

Phylum Ochrophyta, Class Phaeophyceae, Order Dictyotales, Family Dictyotaceae

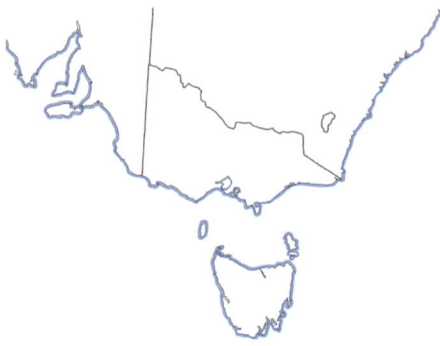

Range
Tropical and/or temperate waters of Australia and elsewhere

Appearance
There are three genera of brown algae in the Order Dictyotales that are broadly similar in morphology and form fan-like thalli: *Padina*, *Zonaria* and *Lobophora*. Growth in these genera arises from apical rows of cells, which can often be seen as paler tips or outer margins. *P. fraseri* grows as clumps of broad fan-shaped fronds with concentric, lightly calcified bands. The outer edge is rolled over and sometimes split. Fronds are three cells thick and grow to 120 mm in height (but tend to be stunted intertidally). There are several species of *Zonaria* in south-eastern Australia. When young, *Zonaria* species are difficult to distinguish from each other and from *Lophophora* spp., with small (20–40 mm), broad, overlapping fan-like fronds. Morphological discrimination of species can become easier as plants grow. For example, *Z. angustata* and *Z. spiralis* both grow to approximately 15 cm in length and have narrow (1–3 mm wide) fronds that are twisted either irregularly (*Z. angustata*) or regularly (*Z. spiralis*). All these species may vary in colour from golden brown to olive green. The gametophytes and sporophytes appear similar and can only be distinguished when fertile.

Habitat and ecology
Padina is generally considered to be a tropical genus, with *P. fraseri* being the only species found on southern Australian shores. *P. fraseri* occurs in the low intertidal zone, in rock pools and on shallow subtidal reefs. *Zonaria* spp. can be found in the low intertidal zone and shallow subtidal zone of rocky reefs, and in rock pools in temperate regions. *Lobophora* spp. can be found from the low intertidal zone of rocky reefs down to a depth of 36 m in warmer temperate waters of south-eastern Australia.

The soft apical growing margins of *Zonaria* spp. (and most likely the other species also) are often grazed by small crustaceans, such as amphipods. Extracts from several species of *Zonaria* have been shown to exhibit strong antibacterial and cytotoxic activity, suggesting potential for pharmacological purposes.

  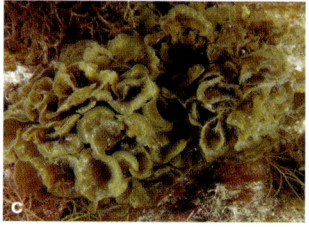

(a) *Padina fraseri*, (b) *Zonaria spiralis* and (c) *Lobophora sonderi* (photograph b by D. Squire, c by J. Huisman).

Neptune's necklace, bubble weed
Hormosira banksii
Phylum Ochrophyta, Class Phaeophyceae, Order Fucales, Family Hormosiraceae

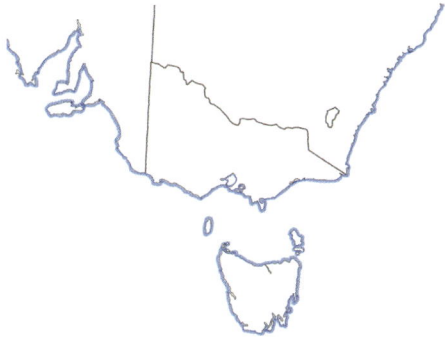

Range
Endemic to Australasia: from Albany, WA, to Lennox Head, NSW, Tas., New Zealand and some smaller offshore islands in southern Australasia

Appearance
The typically olive green fronds of *H. banksii* comprise strings of bead-like vesicles joined together by short stalks (inspiring the name 'Neptune's necklace'). Multiple fronds arise from a discoid holdfast usually attached to the rock (or other hard substrata). Populations vary in morphology, likely related to both environmental conditions and genetic variation. Fronds may be branched or unbranched and vary significantly in length (up to 30 cm long) among populations. Vesicles can also vary greatly in shape (spherical, ovoid, cylindrical, triangular and rectangular) and size (3–30 mm in diameter), with large golf ball-like vesicles found in some estuarine populations. The vesicles are spotted with conceptacles (reproductive structures) from which short white hairs can sometimes be seen at low tide.

Habitat and ecology
Hormosira is the dominant habitat-forming seaweed on most intertidal rocky shores in southern Australia (more sparse with high wave exposure), and is important in facilitating biodiversity, with no functional equivalents. It is vulnerable to anthropogenic disturbances (e.g. trampling, sewage, climate change) and has poor dispersal, making recovery and adaptation difficult. The brown epiphyte *Notheia anomala* is

The morphology of *Hormosira banksii* can vary between sites, and some populations can be heavily epiphytised by *Notheia anomala*, which only grows attached to *H. banksii* (photographs c, d by D. Squire).

often found growing on the plant's surface. The fluid-filled vesicles help resist desiccation at low tide, and they have phenolic compounds in their cell walls that provide protection against UV.

Bull kelp

Durvillaea potatorum, Durvillaea amatheiae
Phylum Ochrophyta, Class Phaeophyceae, Order Fucales, Family Durvilleaceae

Range for *D. potatorum* (blue line) and *D. amatheiae* (aqua line).

Range
D. potatorum: Cape Jaffa, SA, to Bermagui, NSW, Tas. west, south and east coasts, Bass Strait islands

D. amatheiae: Wilsons Promontory, Vic., to Tathra, NSW, Tas. east and north-west coasts

Appearance
Only one species of *Durvillaea*, *D. potatorum*, was recognised in Australia until 2017, when a new, partially co-occurring and morphologically similar species, *D. amatheiae*, was described with the aid of genetic analyses. Both species are referred to as bull kelp. They have tan to dark brown, thick, rubbery blades that vary in width to approximately 300 mm but grow longer in *D. potatorum* (to 8 m) than in *D. amatheiae* (to 5 m); a relatively cylindrical robust stipe (5–50 cm long and 2–12 cm in diameter), usually narrower and shorter in *D. amatheiae*; and a smooth, discoid to bell-shaped holdfast (50–250 mm in diameter) attached firmly to rock. The fronds are solid, non-buoyant and strong, but flexible. *D. amatheiae* fronds are more branched and bear many small lateral blades branching from the under surface on short stalks.

Habitat and ecology
The morphology of bull kelp allows it to withstand intense and frequent wave disturbance. It is most abundant in areas where strong waves break on the rocky reef, occurring at depths of 0–10 m. Unique assemblages of organisms are found in the understorey that can survive the combined scouring effects of wave action and whiplash by the heavy fronds. Thalli are often dislodged and washed up on beaches after storms. The strength and flexibility of the fronds are largely due to high amounts of gelling alginates in the cell walls, for which beach-cast plants are commercially harvested on King Island and the west coast of Tasmania for use in a range of products, including ice cream, toothpaste and fertilisers.

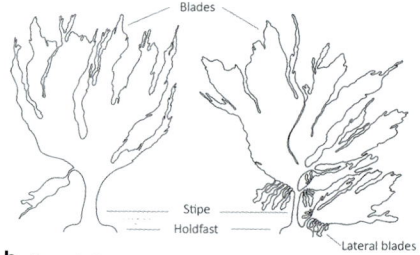

a   **b** *D. potatorum*   *D. amatheiae*

(a) *Durvillaea potatorum* exposed on a very low tide, with large strong holdfast and stipe (inset) to withstand strong wave action. (b) Drawing illustrates differences between *D. potatorum* and *D. amatheiae* (illustrations by X. Keighley).

Crayweed, strap weed
Phyllospora comosa
Phylum Ochrophyta, Class Phaeophyceae, Order Fucales, Family Seirococcaceae

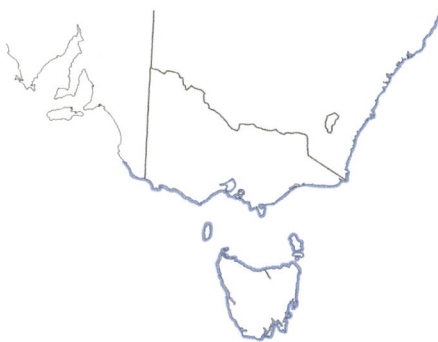

Range
Robe, SA, to Port Macquarie, NSW, and Tas.

Appearance
This olive–brown kelp, often tinted yellow, grows up to 3 m in length. It has a distinct inverted conical holdfast (to 12 cm in diameter) with tightly appressed branches (haptera) that are root-like in appearance. The stipe arises from a concave depression in the holdfast and then flattens, extending into branches with flattened central axes (6–12 mm wide), from which relatively short (usually 5–15 cm) strap-like blades that have toothed margins arise laterally. Often highly branched. Large (to 50 mm long), ellipsoid air bladders occur at the end of short stalks between lateral blades, and only ever have short blades arising from them, if any (c.f. *Macrocystis pyrifera*, p. 48). When reproductive, laterals will appear spotty, with female plants having dark spots and males white; these are the conceptacles that produce eggs and sperm, respectively.

Habitat and ecology
On the mainland, *P. comosa* is abundant on wave-exposed rocky coasts in the shallow subtidal zone, deep rock pools and channels; however, in Tasmania, it can be found in deeper water beyond 18 m. It creates important habitat for many species, including commercially important southern rock lobster (*Jasus edwardsii*), from which the common name crayweed is derived. It is an important food source for abalone. As with most fucoids, it is vulnerable to excessive nutrients and is often lost on coasts where sewage wastewater is discharged. Operation Crayweed is successfully restoring this species to the Sydney coast previously affected by excessive pollution. Endemic to Australia. Currently being explored for aquaculture with a range of potential applications.

Phyllospora comosa (photographs by D. Squire).

Zigzag weeds
Cystophora spp.

Phylum Ochrophyta, Class Phaeophyceae, Order Fucales, Family Sargassaceae

Range
WA, SA, Vic., NSW, Tas.

Appearance
This genus is characterised by a zigzag growth pattern of stipes and branch axes. *Cystophora* holdfasts are discoid–conical and usually small (<3 cm in diameter), even for species with long thalli (to 2 m). Consequently, they have a relatively weak attachment to the substratum and often still have their holdfast attached when washed up on the beach after storms. Most species are monoecious (hermaphroditic). Swollen reproductive receptacles develop on the ends of ramuli, with eggs and sperm developing in bisexual conceptacles that appear as spots on the receptacles. Some species possess air bladders that appear to vary in abundance with environmental conditions, such as wave action. Some species can be easily identified from photographs, but others require closer examination.

Habitat and ecology
Cystophora is the largest genus within the Order Fucales, with 27 species currently accepted. All but one species of dubious taxonomy (*C. fibrosa*) are endemic to southern Australia (23 species) and New Zealand (6 species). *Cystophora* commonly occur in mixed assemblages in the shallow subtidal zone, rock pools and channels. They create important habitat, promoting biodiversity.

Examples of easily identified and common *Cystophora* sp. in south-eastern Australia

Forked zigzag weed
Cystophora retorta

Cystophora retorta (photograph by D. Squire).

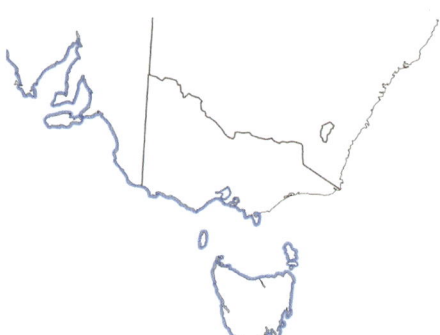

Range
WA to Wilsons Promontory, Vic., and Tas.

Usually light to medium brown, grows to 1.2 m long with a largely two-dimensional structure. Regular, forked, thin (usually 1–5 mm in diameter) branching occurs off relatively straight central axes, with forks distinctively curved at the base. Air bladders, if present, are usually sparse, approximately spherical and small (~10 mm in diameter or less). Thin terminal branches grow to 80 mm in length. *C. retorta* is common in large rock pools, channels and the upper subtidal zone on moderately wave-exposed rocky reefs. Recorded to a depth of 21 m.

Club-leafed zigzag weed
Cystophora torulosa

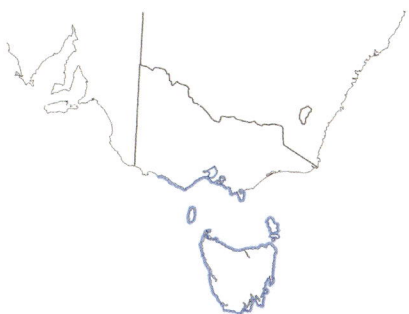

Range
South-west Vic. to Wilsons Promontory, Vic., Bass Strait islands, Tas. and New Zealand

C. torulosa is characterised by clusters of branches arising around the central axis that have cylindrical laterals that are usually forked at the base and have contractions or swellings at intervals along their length. This habit gives the thallus a dense, bushy appearance that provides habitat for epiphytes and epifauna. Thalli can grow up to 1.5 m long and are usually golden to medium brown. Little is known of the ecology and reproductive biology of this species, but it has an interesting biochemical profile. It occurs from low intertidal to upper subtidal zones on moderately wave-exposed rocky reefs.

Cystophora torulosa.

Flat-lobed zigzag weed
Cystophora platylobium

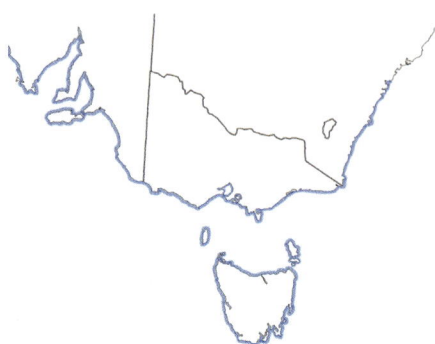

Cystophora platylobium (photograph by D. Squire).

Range
Eucla, SA, to Bondi, NSW, and Tas.

Very distinctive flattened lobe or leaf-shaped blades (10–15 cm long) on short stalks branch off secondary, flattened and weakly zigzagged central axes; zigzagging can be more pronounced at the base, where secondary axes arise from the sides of the primary axis. The broad (to 1.5 cm) primary central axis is flat and undulating. Air bladders are relatively large for this genus (5–15 mm) and primarily spherical on short stalks, but commonly sparse. *C. platylobium* commonly grows to 1–2 m in length in deep waters (10–48 m) on wave-exposed mainland coasts, but in shallower (to 1 m depth) waters in south-east Tasmania. Dark brown (black when exposed to high UV after being washed up on the beach).

Beaded zigzag weed
Cystophora moniliformis

Cystophora moniliformis (photograph a by J. Huisman, b by J. Pocklington).

Range
Cape Naturaliste, WA, to Port Stephens, NSW, Tas. and Lord Howe Island

This species has a distinctive flattened central axis (to 20 mm wide) with only slight zigzagging, from which thin, strongly zigzagged branches arise alternately from the sides, naked near the base, giving a bead-like structure from which the name is derived. Thin (0.2–0.5 mm in diameter) tufted laterals arise from the ends of secondary axes. Lacks air bladders. Usually chocolate brown, but sometimes lighter near the tips; grows to 4 m in length. Common at depths of 1–4 m depth (extending to 28 m) on wave-exposed coasts, but restricted to rock pools where wave action is very strong.

Sargassum weeds
Sargassum spp., *Phyllotricha* spp.
Phylum Ochrophyta, Class Phaeophyceae, Order Fucales, Family Sargassaceae

Range
Widespread across tropical and temperate regions of Australia and globally, with more restricted distributions of individual species

Appearance
The sargassum group is one of the most species-rich and morphologically complex groups of brown algae. Consequently, the taxonomy is controversial and species names are still changing regularly (often informed by genetic data), including *Phyllotricha* being elevated to a genus rather than a subgenus within *Sargassum* in 2012. Consequently, many species within this group can be very difficult to identify, especially from photographs alone. Sargassum weeds can be confused with *Cystophora* spp. in that some have slightly zigzagged stipes, similarly small holdfasts, lengths up to 2 m or more and air bladders. However, sargassum weeds usually have distinctive blades that resemble the leaves of flowering plants with a midrib, usually in addition to blades that look more like seaweeds. Many species within this group have seasonal growth of the upper part of the thallus, whereas the basal leaves and short stalk are present year-round.

Habitat and ecology
Some species have many air bladders and can survive detached and floating on the ocean, often forming very large floating rafts. Sargassum weeds can create important habitat for many

Perennial basal leaves (a) and full dimorphous thallus (b) of *Sargassum* spp. (photographs by D. Squire).

species, whether attached to the substratum or floating. In fact, there have been many documented cases of invertebrates and other algae spreading via floating *Sargassum* rafts. Some species of *Sargassum* (e.g. *S. fusiforme*, known as hijiki in Japan) are both wild-harvested and produced by aquaculture as food.

Golden kelp, leather kelp
Ecklonia radiata
Phylum Ochrophyta, Class Phaeophyceae, Order Laminariales, Family Lessoniaceae

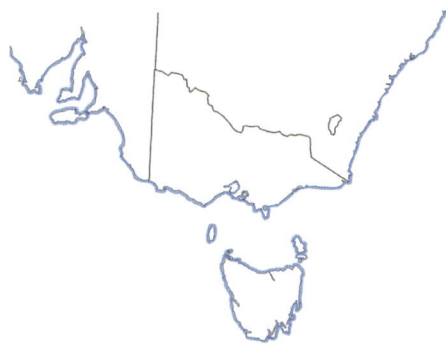

Range
Kalbarri, WA, south, to Caloundra, Qld, Tas., New Zealand and South Africa

Appearance
Golden to dark brown sporophyte grows to approximately 2 m in length with a microscopic gametophyte stage. This kelp is easily recognised with a branched or root-like holdfast (hapterous) that is 5–15 cm in diameter, from which a single, cylindrical and solid stipe (2–12 mm in diameter) arises, with blades consisted of a broad (up to 10 cm wide), smooth central lamina from which thinner lateral fronds (up to 40 cm long) branch off on either side. The lateral fronds are usually corrugated to some degree and spiny. Stipes can vary in length from 2 cm to 1 m, and the central blade can also grow to 1 m in length. New growth begins at the base of the central lamina.

Habitat and ecology
A common kelp throughout the Great Southern Reef on sheltered to moderately wave-exposed shores, in rock pools, channels and at depths to 60 m. This species creates important habitat, sometimes alone and other times mixed with *Phyllospora comosa*, *Cystophora* and/or *Sargassum* spp. (pp. 41–46). Holdfasts of golden kelp and giant kelp (p. 48) also create habitat for diverse

Ecklonia radiata (photograph a by M. Wells, b by D. Squire).

communities. Stipe and blade morphology (e.g. corrugation and spines) vary geographically, influenced by wave action, genetics and other environmental factors. The sporophyte is perennial and found year-round. *E. radiata* is harvested from beach wrack in southern New South Wales for food and is under development for commercial aquaculture in south-eastern Australia, with a range of end uses being examined. *E. radiata* was severely affected by an extreme marine heat wave in Western Australia in 2011, with localised extinction and ecosystem collapse across 100 km of its northern extent, and is likely vulnerable to the ongoing effects of climate change.

Giant kelp
Macrocystis pyrifera
Phylum Ochrophyta, Class Phaeophyceae, Order Laminariales, Family Laminariaceae

Range
SA, Vic. and Tas., New Zealand, Pacific coast of North and South America, South Africa and Subantarctic Islands

Appearance
The large (10–50 cm in diameter and height) conical holdfast of this kelp comprises a tangled mass of intertwining branches (haptera). The holdfast supports several long (to 10 m), thin (<10 mm in diameter), cylindrical stipes that bear flattened blades on one side, each with a large (4–12 cm long, 1–4 cm in diameter) elongate air bladder at its base when mature (c.f. *Phyllospora comosa*, p. 41). Blades are long (from 30 cm to 1.5 m) and narrow (5–15 cm), smooth to corrugated with toothed margins. Giant kelp grows up to 45 m long, but is usually much shorter in Australia. Previously described as *M. angustifolia* in Australia, but molecular analyses revealed a single species of *Macrocystis* with global distribution.

Macrocystis holdfast (b) and blade and air bladders (a; photograph by J. Pocklington).

Habitat and ecology
Giant kelp is found in large offshore beds in areas of strong horizontal water movement and water temperatures <20°C, but smaller plants can occur in rock pools, channels and the subtidal fringe zone. Dramatic declines in abundance and shifts to urchin barrens have been documented in Tasmania due to the combined effects of climate change and overfishing of urchin predators. The Giant Kelp Marine Forests of South East Australia are listed nationally as an endangered ecological community under the *Environment Protection and Biodiversity Conservation Act* in Australia. A rapid growth rate (up to 30 cm day^{-1}) supports wild harvesting in California. Aquaculture production has begun in Chile and is under development in south-eastern Australia. The harvest is used for gelling alginates (see bull kelp, p. 40), feed for abalone aquaculture and food production.

Seagrasses
Phylum Magnoliophyta (Angiosperms), Class Liliopsida, Order Alismatales

Appearance
Seagrasses are true plants, not algae, with a vascular system for transporting water and nutrients from the roots to the leaves, as well as flowers for sexual reproduction. In most seagrasses, flowers and fruits are inconspicuous, with water-borne pollination and dispersal. Seagrasses have creeping horizontal rhizomes and root systems (like many terrestrial grasses), multiple stems arising from the rhizomes and green leaves, with veins running throughout.

Habitat and ecology
Seagrasses occur in intertidal and subtidal environments. Flowering plants evolved on land, but some lineages have become secondarily marine, requiring adaptations to cope with high salinity, water-logged sediments and low light and dissolved CO_2 for photosynthesis during submersion. Seagrass leaves can also become covered with small animals and algae (epiphytes), especially in wave-sheltered environments, causing additional light limitation. Seagrasses often form extensive beds, spreading by rhizomal growth, and create important habitats for many species and nursery grounds for juvenile fish. Seagrass leaves often wash up on beaches *en masse* and sometimes aggregate into distinctive balls. Dried seagrasses were historically used as insulation in housing in Australia until it was realised this practice was unsustainable. There are around 30 species of seagrass in a single order (Alismatales) in tropical and temperate Australia, but only four key groups in south-eastern Australia.

Seagrass balls (photograph by D. Squire).

Eelgrass, garweed
Zostera spp.
Family Zosteraceae

Zostera nigracaulis (photograph by P. Macreadie).

The taxonomy of this group of seagrasses is controversial, but genetic analysis most strongly supports *Heterozostera* as a subgenus of *Zostera* (adopted here).

Range
Species-specific, with some overlap in south-eastern Australia

Zostera muelleri: Yorke Peninsula, SA, to Sussex Inlet, NSW, Tas., New Zealand

Zostera nigracaulis: Dongara, WA, to SA, Vic., NSW, north and east Tas.

Zostera tasmanica: Portland to Wilsons Promontory, Vic., north and east Tas.

Zostera polychlamys: Dongara, WA, to Investigator Strait, SA

Eelgrasses have long, thin (1–2 mm wide), yellow-green to bright green leaves with longitudinal veins that encircle the stems forming a sheath. Flowers develop on a flattened spike (spadix) within a modified leaf sheath (spathe). Rhizomes have distinct nodes from which unbranched roots arise. *Z. nigracaulis* can

Zostera polychlamys (photograph by J. Huisman).

be distinguished by its black wiry stems and *Z. polychlamys* by the absence of erect stems, but reliably distinguishing other species based on morphology usually involves microscopic examination. However, *Z. muelleri* usually occurs in sheltered estuaries, whereas *Z. nigracaulis*, *Z. tasmanica* and *Z. polychlamys* are usually found in coastal waters from rock pools to depths of 8–15 m, although *Z. polychlamys* can occur deeper, to at least 20 m. Seasonal flowering and fruiting times vary among species.

Paddleweed
Halophila australis, Halophila ovalis
Family Hydrocharitaceae

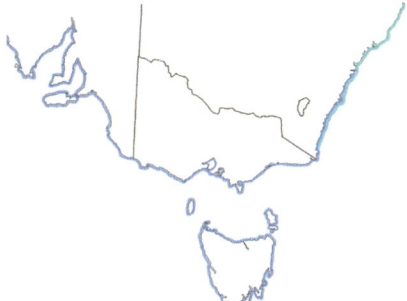

Range for *H. australis* (blue line) and *H. ovalis* (aqua line).

Halophila australis (photograph by J. Huisman).

Halophila ovalis (photograph by J. Huisman).

Range
H. australis: Dongara, WA, to Sydney, NSW, Tas.

H. ovalis: north of Cowarump Bay, WA, to near Eden, NSW

Although several species are tropical, the two southern Australian paddleweeds are easily recognised by their distinctive ovate, paddle-shaped leaves (with an obvious mid-vein and lateral veins) that occur in pairs on branched stems. The ranges of these species overlap in Western Australia and New South Wales, where they may be most easily distinguished (in the absence of flowers) by the average proportional lengths of the leaves: *H. australis* leaves are longer (to 70 mm) with a length/breadth ratio of 3–4, whereas *H. ovalis* leaves have a length/breadth ratio of 2–3 and a maximum leaf length 40 mm.
H. ovalis is found intertidally on sand and mud in coastal and estuarine environments, whereas *H. australis* occurs in wave-sheltered soft sediments at depths of 0–23 m.

Tapeweed
Posidonia spp.
Family Posidoniaceae

Posidonia sp. (photograph by P. Carnell).

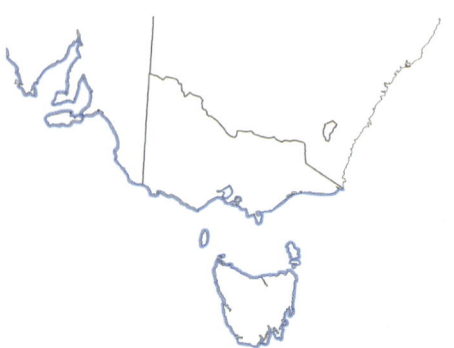

Range
WA, SA, Vic., Tas.

There are several species of *Posidonia* with different ranges across southern Australia.

Tapeweeds are distinctly grass-like but have broader (usually flattened) leaves (typically >10 mm wide) than eelgrass. Like eelgrasses, roots (one to two) arise from the nodes along rhizomes, but in tapeweeds these alternate between being branched and unbranched. Tapeweeds also have distinctive spongy fruits that contain a single seed. Without microscopic examination, the width of the leaves and the distribution information is most helpful for discriminating species.

Sea nymph
Amphibolis antarctica
Family Cymodoceaceae

Amphibolis antarctica (photograph by P. Carnell).

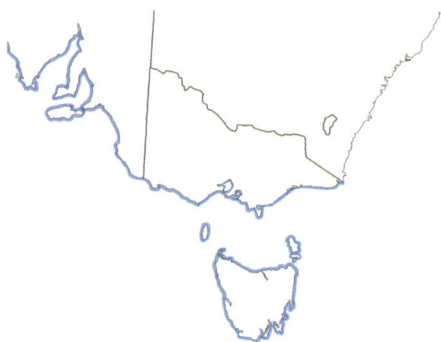

Range
WA, SA, Vic, Tas.

The stems of this seagrass are long, thin and woody. Distinctive clusters of rectangular, twisted, moderately dark green leaves up to 50 mm long grow from the tips of the stems. Sea nymph occurs in rock pools and shallow subtidal areas on shores of moderate to low wave exposure and can be a significant component of beach-cast wrack on some shores. Seeds germinate on the parent plant, and seedlings detach after 7–12 months, bearing distinctive hooks for snagging onto other plants or the sea floor.

Saltmarsh plants
Phylum Magnoliophyta (Angiosperms)

Appearance
Saltmarsh plants are vascular, flowering plants that are secondarily marine; there are many lineages of plants that evolved on land and then reinvaded tidal marsh habitats. There are broadly three types of saltmarsh plants, namely succulents, grasses and herbs, each with different solutions to the challenges of tidal inundation and different associated morphologies. All species have true roots, stems and leaves, as well as woody tissues to provide structural support during tidal emersion.

Habitat and ecology
Saltmarshes are defined by the presence of these types of plants and periodic tidal inundation. In most cases they are associated with estuaries. Species are typically distributed in distinct zones from the low- to high-tide mark. Saltmarshes (as well as seagrass and mangrove habitats) have historically been undervalued in Australia and elsewhere, and often degraded. However, in addition to their various ecological functions and provision of biodiversity, we now understand these coastal vegetated habitats are extremely important for capturing and sequestering carbon, such that restoration of these habitats should be prioritised to assist in climate mitigation.

Succulents
Beaded glasswort, beaded samphire, sea asparagus
Sarcocornia quinqueflora
Family Chenopodiaceae

Range
Carnarvon, WA, to SA, Vic., NSW, Qld, Tas., New Zealand

Beaded glasswort grows as a dense turf of perennial, cylindrical, segmented (5- to 15-mm-long segments), succulent stems that spread horizontally and become erect at the end to a height of approximately 50 cm. Glassworts appear leafless, but scale-like leaves grow opposite each other and around the stem, joining at the base and forming the joints. Roots form at nodes of horizontal stems. Inconspicuous, segmented flower spikes (20–50 mm long) occur at the end of the stems most of the year. This

Sarcocornia quinqueflora (photograph by D. Squire).

species is abundant in the lower saltmarsh where tidal inundation is frequent, whereas the related species *Sarcocornia blackiana* usually occurs higher where inundation is irregular. The soda-rich ash of burnt glasswort was historically used to make soap and glass (from which this name was derived).

Shrubby glasswort, samphire, sea asparagus
Tecticornia arbuscula
Family Chenopodiaceae

Range
South-eastern WA to NSW, Tas.

Tecticornia spp. are morphologically similar to *Sarcocornia* spp. but tend to have more woody basal stems that branch into the fleshy succulent, cylindrical, segmented stems and form discrete shrubs rather than a ground cover. There are also subtle differences in flowers and seeds between genera. *T. arbuscula* is common in coastal saltmarsh across southern Australia and forms a highly branched shrub to small tree (to 2.5 m tall). Samphire/sea asparagus (both *Sarcocornia* and *Tecticornia* spp.) are nutritious edibles (with high vitamin and mineral contents) that have been important foods to Aboriginal peoples for millennia, and are commercially available in some restaurants. Generally, the young, fleshy stems are consumed.

Tecticornia arbuscula forms a small shrub with woody stems and branches in the background, whereas *Sarcocornia quinqueflora* creates a spreading dense turf in the foreground (a). The fleshy, succulent, cylindrical, segmented stems of both species are similar in appearance (b, close-up of *T. arbuscula*) and can vary from yellow to red in colour. (Photograph b by D. Squire.)

Austral seablite
Suaeda australis
Family Chenopodiaceae

Range
Abrolhos Islands, WA, south, to Townsville, Qld, Tas.; sometimes inland in wet saline areas

Small, rounded perennial shrub to 70–80 cm tall, with branching occurring from the base. Leaves are succulent, light to medium green or reddish-purple, rounded on one side and flat on the other, and can be slender to 30 mm long or fusiform/spindle-shaped to approximately 10 mm long. Stems green to rich red. Minute, greenish flowers are present at the base of the leaves where they join the stem, usually from spring to autumn but year-round in some areas. After fertilisation, small (1.5–2 mm in diameter), succulent, button-like fruits with five lobes form in the leaf axes. Grows in mudflats and estuarine saltmarsh. Edible traditional food for Aboriginal peoples, commercially harvested or cultivated in some areas and can be used steamed, stir-fried or blanched.

Suaeda australis.

Grasses
Australian saltgrass
Distichlis distichophylla
Family Poaceae

Range
SA to Narooma, NSW, Tas.; isolated records from Esperance, WA

This perennial grass has a creeping habit, spreading via rope-like rhizomes. Leaves are thin, rough and spiky, alternating on opposite sides of the stem. It can form prickly clumps to 30 cm tall. Separate male and female plants produce morphologically different flowers (usually 0.5–4 cm long) from October to April in Victoria and South Australia, and in spring or after rain in New South Wales. Seed heads develop as compact arrowhead spikelets with overlapping scales.

Distichlis distichophylla.

Rhizome and root systems are important for stabilising sediments in saltmarshes and dune systems.

Common reed
Phragmites australis
Family Poaceae

Range
Coastally Adelaide, SA, to far-north Qld, north and east coast Tas., a few coastal records for WA and NT; also inland in all states, New Zealand

This is a tall (1–3 m), fast-growing perennial grass that is common in the low shore (often partially submerged) in estuaries and swampy wetlands. Stems are cane-like and grey-green leaves are broad and long, tapering to a fine point. Flowers are large (to 40 cm long), hairy plumes, beige to purplish in colour and numerous on the ends of canes from November to May, depending on region. This species provides important habitat for vertebrates and invertebrates and is a buffer against tidal and downstream erosion of estuary banks. Roots are a traditional food of some Aboriginal peoples.

Phragmites australis.

Herbs
Creeping brookweed
Samolus repens
Family Primulaceae

Range
Port Hedland, WA, south, to near Brisbane, Qld, Tas., New Zealand

This plant produces creeping horizontal stems for vegetative spread and vertical stems that can grow to 30 cm tall. The basal leaves are usually thick, slightly pear-shaped and dull green, whereas leaves on the erect stems are narrower and more elongate, tapering at the base. Small (10 mm in diameter), white (or pink) star-shaped flowers are produced in terminal clusters on erect stems from September to February. Common in saltmarsh as well as primary dunes, saline grassy wetlands and woodland swamps. Foraging habitat for birds and aquatic insects.

Samolus repens.

Grey mangrove
Avicennia marina
Phylum Magnoliophyta (Angiosperms), Class Magnoliopsida, Order Lamiales, Family Acanthaceae

Range
Mainly a tropical species, but extends as far south as Bunbury on Australia's west coast, Jervis Bay on the east coast, in South Australia (Gulf of St Vincent, Spencer Gulf and the west coast of Eyre Peninsula) and in Victoria (between Barwon Heads and Wilsons Promontory; the most southern location for mangroves in the world)

Appearance
Mangroves are distinguished by their complex, but shallow, root systems that include anchoring roots that extend laterally and provide horizontal support, aerial prop roots that extend from the trunk downwards for support, vertical aerial roots (known as pneumatophores; up to 20 cm long and 1 cm in diameter in this species) and fine roots for nutrient uptake. Grey mangrove is the most common mangrove in south-eastern Australia and the only species that grows south of New South Wales. It grows as a small shrub or tree to 14 m in height, with generally shorter heights in southern latitudes. The bark, under surface of leaves (generally ovate to elliptical and narrower at the base; 3.5–12 cm long, 1–4 cm wide) and the closed flower heads are silvery grey, whereas the upper surface of leaves is smooth and glossy green. White to golden flowers (<1 cm in diameter) occur in clusters of three to five and are present most of the year.

Habitat and ecology
Grey mangroves grow in the waterlogged, intertidal sandy and muddy sediments of estuaries and tidal marshes. The root systems are submerged at high tide, but at low tide the vertical roots gather oxygen. The root systems provide important habitat for invertebrates and algae, as well as nursery grounds for many fish species. Grey mangroves have high salt tolerance and excrete salt from glands in their leaves.

Avicennia marina (photograph by P. Carnell).

Lichens

Verrucaria spp., *Lichina* spp., *Caloplaca* spp.
Phylum Ascomycetes

Range
Verrucaria spp.: SA, Vic., Tas.

Lichina spp.: Vic., southern NSW, northern Tas.

Caloplaca spp.: global

Appearance
Verrucaria spp. occur in bands of thick, black mats. *Lichina* spp. are smaller, black, tufted lichens that occur in patches and never in bands. *Caloplaca* spp. are orange lichens that often occur as leafy encrustations.

Habitat and ecology
Lichens are a symbiotic association of an alga or cyanobacterium and a fungus. Most rocky shore lichens are found above the intertidal zone (in the splash zone) on shores that experience moderate to strong wave exposure. *Lichina* also occurs with, and is a food source for, periwinkles in the upper shore zone on many southern shores.

(a) *Lichina* and (b) *Caloplaca*.

Animals of rocky shores

Introduction

The structure of rocky shores will depend on the local geology. Common rocky shore types are wave-cut platforms, large granite blocks, basalt flows, boulders and loose rocks, as well as mixed shores of eroded platform with rock rubble. The plants and animals encountered on these shores in south-eastern Australia will be very similar regardless of shore type. However, shore topography, wave exposure, sand scouring and other local-level factors may influence which species are more commonly encountered on any particular shore.

Wave-cut platforms are formed by the erosion of sedimentary rock (limestone, sandstone or mudstone) and are often backed by cliffs. Rock type, tides, waves and weathering combine to influence surface topography. Limestone can have lots of sharp jagged projections, whereas sandstone and mudstone tend to be smoother. Expanses of Neptune's necklace *Hormosira banksii* (p. 39) and rock pools lined with algae are typically encountered on these platform shores. The more obvious animals living among the algae and on patches of rock include periwinkles (pp. 98–99), limpets (pp. 86–88), top shells (pp. 92–93) and whelks (p. 102).

The most conspicuous plants and animals on **granite shores**, such as those along the north-east coast of Tasmania up to Wilsons Promontory, are animals that attach themselves firmly to the rock. Lichens (p. 58), barnacles (pp. 118–119), mussels (pp. 110–111), *Galeolaria* (p. 74) and limpets (pp. 86–88) often occur in distinct bands where the granite slopes steeply to the waterline.

The top surface of dark **basalt** rock may become too hot between tides to support life, with most animals and plants found on vertical surfaces, in crevices and in small rock pools. Here the black nerite (p. 96), striped-mouth conniwink (p. 97), barnacles (pp. 117–121) and encrustations of *Galeolaria* (p. 74) are commonly encountered.

Shores comprising large **mudstone** boulders and platforms are typical of the coastline between Lorne and Cape Otway in Victoria, where wave exposure is a factor influencing species distribution. Algae are less obvious on these shores, except close to the low tide level, where coralline algae and bull kelp dominate.

Platform shore covered in a dense mat of *Hormosira banksii* (page 39).

Basalt boulder shore.

Rock size and wave exposure influence species diversity on shores comprising **boulders and rock**. On coastlines with high wave exposure, smaller rocks (cobbles, stones and pebbles) are constantly moved over each other with each breaking wave, forming an unstable environment for animals and plants. More sheltered shores, and those comprising larger, more stable rocks and boulders, house a fauna of which molluscs, barnacles and crabs are the most obvious and abundant members. Many of the animals living on the exposed surfaces and in crevices between rocks are also found on shore platforms, whereas crabs (pp. 133–140, 142) shelter underneath the rocks. Also living under rocks are animals that tend to avoid light, moving in search of shade whenever rocks are overturned (e.g. chitons, pp. 79–81).

Shores comprising a mix of **eroded platform with rock rubble** have habitat for animals that live on rock surfaces, as well as those that seek shelter under rocks. These shores have the potential for high faunal diversity and may reward the avid naturalist with some more cryptic finds, such as flatworms (p. 69), peanut worms (p. 78), keyhole limpets (p. 84), ribbon worms (p. 77), chitons (pp. 79–81) and brittle stars (p. 153).

It is not uncommon for rocky shore ramblers to spend much of their time on the shore exploring **rock pools** – holes and depressions eroded into the surface of intertidal rocky shores where salt water is retained during low tide. Rock pools are

Animals of rocky shores

Mudstone boulder shore.

Surface of mudstone platform shore.

Chitons on an overturned rock.

Sediment-covered rocks on a shore with low wave exposure.

Eroded platform shore with rock rubble.

Mid-shore rock pool on a platform shore.

Typical upper intertidal (a) and sublittoral fringe (b) zone communities on a rocky shore.

home to animals and plants that may not otherwise survive exposure to air as the tide recedes. Pools differ in their size, shape, depth, zone on the shore and wave exposure; all these factors influence the type and diversity of species found within the pools. Deep pools close to the water's edge at low tide are windows into the diverse array of plants and animals more typical of the shallow subtidal zone. It is here that you are likely to find many of the seaweeds featured in the Plants section of this guide. Invertebrates you may find in rock pools, but not on the surrounding exposed rock, are nudibranchs (p. 109), hermit crabs (p. 127), seaweed crabs (pp. 129–130), shrimp (p. 126), sea tulips (p. 156) and small fish.

The **reef fringe** (sublittoral fringe zone) may be uncovered during the very low spring tides, often exposing undercut reef, caves and crevices. This fringe zone is home to many seaweeds and colourful encrusting invertebrates such as sponges (p. 63), bryozoans (p. 143) and compound ascidians (p. 157), along with crevice-dwelling animals such as elephant snails (p. 83), abalone (p. 82), red bait crabs (p. 142) and juvenile rock lobster.

Sponges
Phylum Porifera, Class Demospongiae

Examples of more common intertidal sponges in south-eastern Australia
Golf ball sponge *Tethya* sp.

Branching sponge *Callyspongia* sp.

Encrusting sponge *Dendrilla rosea*

Range
Australia wide

Appearance
Sponges are simple animals in that they lack true tissues and organs and live permanently attached to hard surfaces. Golf ball sponges are spherical and up to 50 mm across. They have a tough outer covering that is often a shade of pink or orange. Encrusting forms can be a variety of shapes, textures and colours, but all have a spongy feel when touched. Sponges are difficult to identify from appearance alone, with identification usually requiring microscopic examination of skeletal elements called spicules.

Habitat and ecology
Sponges are found in rock pools, on the under surface of rocks and under ledges in the low intertidal zone and below. The growth form of intertidal sponges may be limited by the disturbance created by waves; for example, flat encrusting forms may grow tall and bulky in underwater areas that are free from waves. Sponges are filter feeders. Unpleasant chemicals and sharp

(a) *Tethya* sp., (b) *Callyspongia* sp. and (c) *Dendrilla rosea*.

spicules in their skeleton deter most predators. Sponges can regenerate from even the smallest fragment and are able to reshape in response to water flow.

Anemones
Phylum Cnidaria, Class Anthozoa

Anemones comprise a hollow cylindrical body (column), with a muscular base (pedal disc) at one end and a mouth surrounded by tentacles at the other (oral disc). The tentacles require water pressure to extend, so are often not visible when anemones are uncovered during low tide. This feature also helps anemones conserve water.

Red waratah
Actinia tenebrosa
Phylum Cnidaria, Class Anthozoa, Family Actiniidae

Actinia tenebrosa with tentacles retracted inside the body (a) and extended (b).

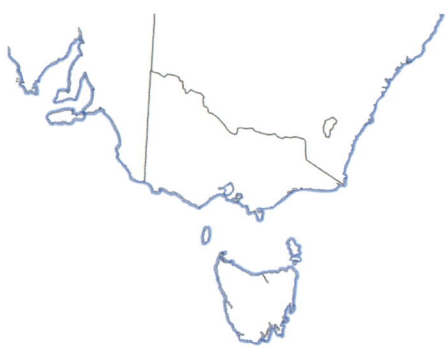

Range
Southern WA, SA, Vic., NSW, Tas.

Appearance
A. tenebrosa often appears as a smooth, shiny, deep red blob with a paler central dot when the light red tentacles are retracted inside the body during low tide. The column is widest at the base, which adheres firmly to the substratum. Individuals may be up to 40 mm in diameter when the column is extended and have numerous short red tentacles.

Habitat and ecology
Individuals of this species are found under ledges, in crevices and in pools in the mid- to upper intertidal zones on moderately exposed shores. Reproduction is either by brooding clonal young that are released onto nearby rock or the release of larvae into the sea for wider dispersal. This species feeds on any small animals or dead material that drift close enough to be captured by stinging cells (cnidocytes) fired from the tentacles.

Green anemone, snakelock anemone
Aulactinia veratra
Previously *Cnidopus verater*

Phylum Cnidaria, Class Anthozoa, Family Actiniidae

Aulactinia veratra.

Range
Southern WA, SA, Vic., NSW, southern Qld, Tas.

Appearance
A. veratra is a common intertidal anemone. The column colour is usually a shade of green (from olive green to dark bottle green), although brown and dull red individuals are sometimes found. The column is characterised by numerous rows of warty growths called verrucae. The oral disc is 40–70 mm in diameter. The tentacles are thin, smooth and up to 40 mm long. The colour of the tentacles is a uniform lighter shade of the column colour.

Habitat and ecology
Individuals of this species may be found in crevices, under overhanging rock and in rock pools in the mid- to low intertidal zones. *A. veratra* may occur in dense aggregations. As with most anemones, this species feeds on small animals captured from the surrounding water using the stinging cells (cnidocytes) on the tentacles.

Shellgrit anemone, speckled anemone
Oulactis muscosa
Phylum Cnidaria, Class Anthozoa, Family Actiniidae

Oulactis mucosa (a) and an undescribed species of *Oulactis* found in Port Phillip Bay (b).

Range
Southern WA, eastern SA, Vic., NSW, southern Qld, Tas.

Appearance
The column of individuals of this species is often buried in sand-filled cavities on the rock surface, with only the oral disc and tentacles visible. The exposed grey–white tentacles may be speckled, striped or blotchy, and are usually covered with particles of sand and broken shells (shell grit). The colour of the oral disc varies and may be red, brown, black, green or white. Individuals grow up to 100 mm tall and 60 mm in diameter.

Habitat and ecology
This common species is found in sand-filled crevices and cavities in the mid- to low intertidal zone on sheltered to moderately exposed rocky shores. It feeds on small mussels that have become dislodged from surrounding mussel beds. As with other anemones, reproduction can be via budding, where fragments break off and develop into new, genetically identical individuals. Reproduction also occurs via the release of eggs and sperm into the surrounding water for fertilisation. The fertilised egg develops into a free-swimming pear-shaped larva (planula) that attaches to a solid substratum to develop into the adult form.

Small brown anemone, mudflat anemone
Anthopleura hermaphroditica
Phylum Cnidaria, Class Anthozoa, Family Actiniidae

Range
SA, Vic., NSW, Tas.

Appearance
This small (<10 mm in diameter), common anemone is often speckled brown or olive green with radiating pale and dark lines on the oral disc. The colour comes from symbiotic algae (zooxanthellae) that live inside the anemone's tissues.

Habitat and ecology
This species is found in the intertidal zone and shallow subtidal areas in sheltered bays and estuaries, on rocks, among tubeworm reefs and attached to cockles or pebbles just beneath the sediment surface. It forms a symbiotic relationship with the bivalves from the genus *Katelysia*, with both the anemone and the bivalve gaining protection from predation through this association. This anemone tolerates changes in salinity and high concentrations of organic matter, attributes that are typical of estuarine environments.

Anthopleura hermaphroditica.

White-striped anemone
Anthothoe albocincta
Phylum Cnidaria, Class Anthozoa, Family Sagartiidae

Range
Southern WA, SA, Vic., NSW, Tas.

Appearance
A. albocincta has an orange-and-white–striped column, orange oral disc and many short, white tentacles. When disturbed, individuals of this species eject white threads loaded with 'stinging cells' (cnidocytes), from holes in the body wall. This is a relatively small anemone, with a disc diameter up to 15 mm.

Habitat and ecology
This species is usually found in clusters under ledges and rocks in the lower intertidal zone, often hanging with the oral surface, in which the mouth is located, pointing downwards. Aggregations of clones are formed when individual adults split in half to form two individuals.

Anthothoe albocincta (photograph b by P. Davis).

Flatworms
Phylum Platyhelminthes

Free-living flatworms have flattened, unsegmented bodies, are bilaterally symmetrical and have only one external opening that serves as both a mouth and an anus. Flatworms 'breathe' via diffusion across their body surface and lack a circulatory system, both of which limit the size to which flatworms can grow.

Southern (common) flatworm
Notoplana australis
Phylum Platyhelminthes, Class Turbellaria, Family Notoplanidae

Notoplana australis.

Range
SA, Vic., NSW, Tas.

Appearance
The body form of *N. australis* is roughly oval and flattened. The soft, flexible body is often so translucent that the branches of the digestive system are clearly visible. The edges of the body may appear crinkled. These flatworms vary in colour from white through grey to brown.

Habitat and ecology
The common flatworm occurs in rock pools, under rocks and in beds of mussels and tubeworms on sheltered shores. The common flatworm preys on living or dead animals by everting part of its gut through its mouth. This flatworm must remain moist to facilitate gas exchange across the body surface but has little protection against drying out, so is only found in areas that remain moist and are sheltered from the sun. Most flatworms have both male and female reproductive organs but still transfer sperm between separate individuals for reproduction. This species has planktonic larvae.

Segmented or bristle worms
Phylum Annelida, Class Polychaeta

Polychaetes are segmented worms with paddle-like flaps, called parapodia, on each segment. Numerous bristles, called chaetae, protrude from the parapodia. Polychaetes are grouped according to whether they are free moving or sedentary (e.g. tube dwelling). Free-moving forms generally have a distinct head end bearing tentacles, a mouth and light-sensitive eyes in some groups. Almost all polychaetes live in marine environments.

Scale worm
Lepidonotus melanogrammus
Phylum Annelida, Class Polychaeta, Family Polynoidae

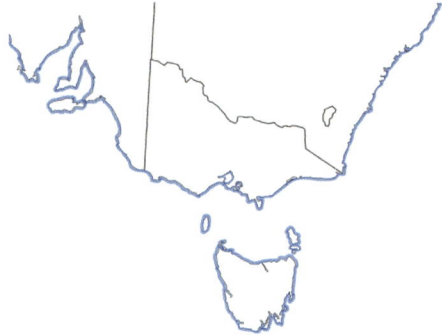

Range
Southern WA, SA, Vic., NSW, Tas.

Appearance
Scale worms have a pair of shiny scales (elytra) covering the upper surface of each segment. They have a well-defined head and relatively few body segments compared with other polychaetes. The genus *Lepidonotus* includes around 80 described species globally. *L. melanogrammus* has 12 pairs of overlapping, rounded scales characterised by purple to brown colouration on the outer margins. Individuals can grow up to 50 mm long and 10 mm wide.

Habitat and ecology
This species is found under rocks, in mussel beds and on kelp holdfasts on shores with moderate to strong wave exposure. *L. melanogrammus* is often found in association with the brittle star *Ophionereis schayeri* (p. 153), although the nature of this association is unknown. Scale worms are slow-moving predators that feed on small animals, capturing prey with an eversible pharynx. They reproduce sexually and release planktonic larvae for dispersal.

Lepidonotus melanogrammus.

Ragworms
Phylum Annelida, Class Polychaeta, Family Nereididae

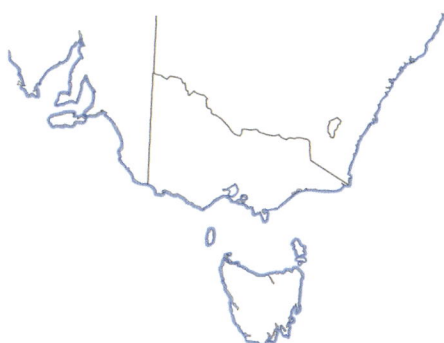

Range
Australia wide

Appearance
Australian nereid polychaetes typically grow up to 100 mm long with up to 150 body segments, although some species of *Perinereis* may grow much longer. The head has four dark eyes, two smooth fleshy palps, one pair of small antennae and four pairs of tentacles. A large proboscis armed with jaws is extended from the head to capture food. Each body segment has noticeable side projections (parapodia or 'false feet').

Habitat and ecology
Nereid polychaetes are common, abundant and widely distributed. Ragworms are found under rocks and in algal mats, mussel beds and encrustations of tubeworms. Ragworms eat both plant and animal matter and actively forage for prey or organic debris. As with most nereid worms, the sexes are separate and reproduction is sexual. The adult worm matures into a modified, fertile form called an epitoke. Thousands of epitokes swarm to the water surface at the same time for the synchronised release of eggs and sperm. Adults usually die after spawning.

(a) Nereid polychaete with head and body appendages visible. (b) Nereid polychaete among worm tubes. (c) The epitoke of a nereid worm.

Terebellid worms, spaghetti worms
Eupolymnia koorangia
Phylum Annelida, Class Polychaeta, Family Terebellidae

Range
Australia wide

Appearance
This worm has a soft, flaccid body up to 50 mm long, with a tangled mass of long tentacles arising from the head. Live individuals are salmon pink in colour with numerous white dots scattered along the sides of the body.

Habitat and ecology
This terebellid worm is found in crevices and under rocks in sheltered areas. It cements sand grains and pieces of debris together to construct a tube-like body covering. Terebellids move their tentacles through the sediments in search of organic matter (food), which is transported along grooves in the tentacles towards the mouth. These constantly moving tentacles are often the only visible sign of the worm's presence. Over 85% of the Australian terebellid species are endemic to Australia.

Eupolymnia koorangia.

Native fan worm
Sabellastarte australiensis
Phylum Annelida, Class Polychaeta, Family Sabellidae

Sabellastarte australiensis.

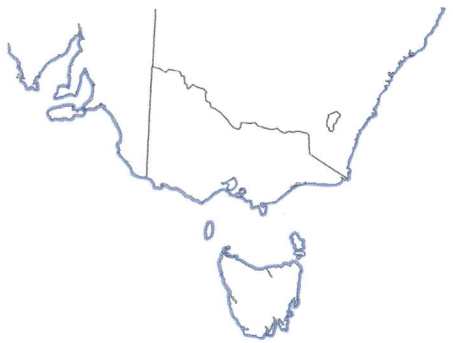

Range
Southern WA, SA, Vic., NSW, southern Qld, Tas.

Appearance
The worm-like body of fan worms is hidden within a tough, leathery mucous tube that can be embedded with mud and sand. The tube opening lacks a covering cap (operculum). A crown or tuft of feathery tentacles emerges from one end of the tube during feeding, but is rapidly withdrawn into the tube when the animal is disturbed. This species grows to 10 mm. The native fan worm has a double circular feeding crown that is a deep purple and appears banded with orange specks, although there is some colour variation along the species range: the fan is white in eastern Victoria and brown–orange in Port Phillip Bay.

Habitat and ecology
Sabellastarte australiensis is found in and among reef rubble, on reefs, rock pools and jetty pylons in sheltered bays and on the exposed coast. Fan worms are active filter/suspension feeders.

Note, the introduced fan worm *Sabella spallanzani* is much larger than the native fan worm and is generally a subtidal species. It can be found in very large numbers, blanketing the sea floor and is likely to have a significant negative impact on native invertebrate species. For example, studies have shown that *S. spallanzani* outcompetes other filter feeders.

Sydney 'coral'
Galeolaria caespitosa
Phylum Annelida, Class Polychaeta, Family Serpulidae

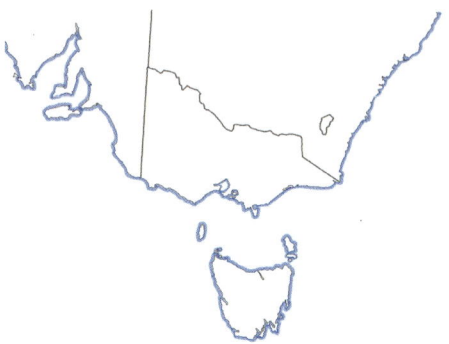

Range
Southern WA, SA, Vic., NSW, southern Qld, Tas.

Appearance
The white calcareous tubes of *G. caespitosa* form a distinctive band on many rocky shores. Tubes may occur alone or in massive encrustations, the latter often found around pier pilings. A moveable cap (operculum) covers the tube opening and the worm's head bears a fan of black, feathery tentacles.

Habitat and ecology
This species is common in the mid- to lower intertidal zones on all shores and on pier pilings. Encrustations of tubes support a diverse assemblage of other invertebrate species, including the small bivalve *Lasaea australis* (typically on the open coast), mussels, ragworms, brittle stars and crustaceans, particularly crabs and amphipods.

Individuals use their tentacles to filter food particles from the water at high tide. Males and females release sex cells (gametes) into the water, where fertilisation occurs. Note that this animal is not a coral (Phylum Cnidaria) but a polychaete worm, despite its common name.

Galeolaria caespitosa: colony of encrusting calcareous tubes that house individual worms (a) and an individual worm outside of its tube (b).

Australian tubeworm, estuarine coral
Ficopomatus enigmaticus
Phylum Annelida, Class Polychaeta, Family Serpulidae

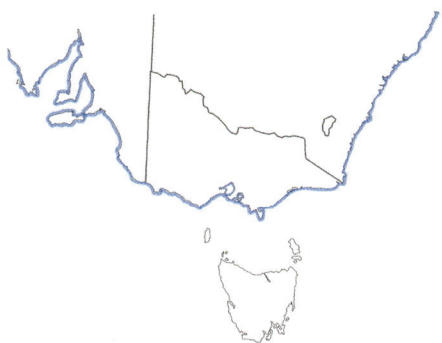

Ficopomatus enigmaticus: colony (a) and individuals out of their tubes (b).

Range
Southern WA, SA, Vic., NSW, Qld

Appearance
F. enigmaticus is a distinctive reef-building tubeworm in southern Australian estuaries, particularly those that intermittently open and close at the estuary mouth. It can establish on any hard surface, from small pebbles to bottle tops and a range of infrastructure (e.g. pier and bridge pylons, pipes and rock walls). Each reef consists of hundreds of small, individual worms that have a cap (operculum) that protects the entrance of their hard, calcareous tube. The worms have three distinct body regions: feeding crown, collar (thorax) and abdomen. This species can be confused with another closely related worm, namely *Galeolaria caespitosa* (previous page). These species are distinguished by differences in their tubes and operculae. They also occur in different habitats, with *Galeolaria* typically found on more wave-exposed coastlines.

Habitat and ecology
As with *Galeolaria*, *Ficopomatus* colonies provide homes for several other invertebrates, such as amarinus crabs, mussels (*Xenostrobus securis*), ragworms and amphipods. The colonies filter large volumes of water, removing considerable amounts of suspended material from the water column. Individuals of this species are often found in the stomach of black bream.

Presumed endemic to Australia, the true natural range for this species is unknown. It typically occurs in temperate estuaries of southern Australia, but has become an invasive pest species in some countries outside of Australia.

Spirorbid worms
Spirorbis sp.
Phylum Annelida, Class Polychaeta, Family Spirorbidae

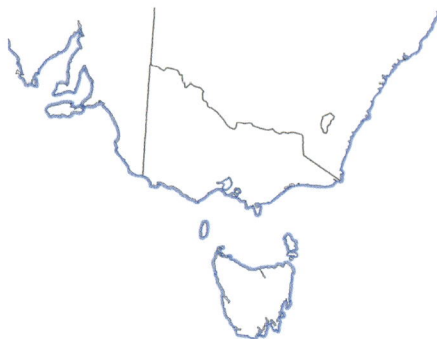

Range
Worldwide

Appearance
Spirorbid worms live in spirally coiled calcareous tubes attached to hard surfaces, seaweed and seagrass. A cap (operculum) is used to close off the end of the tube to prevent the animal inside drying out during low tide and to provide protection from predators. Spirorbid worms are generally less than 10 mm in diameter. The direction of coiling differs between species.

Habitat and ecology
Spirorbid worms are found attached to a range of surfaces, such as rocks, seaweed and the hard shells of other animals. They filter feed using a tentacular fan that protrudes from the shell when the animal is covered with water. The young are incubated in brood chambers inside the tube. The hatched larvae are highly habitat-specific at settlement.

Spirorbis sp. (a) colony and (b) close-up of individuals in a colony.

Ribbon worms, proboscis worms
Phylum Nemertina

Range
Worldwide

Appearance
Nemerteans are worms with a smooth body surface and may be round or flattened. The head end has a ventral mouth through which a muscular proboscis is extended for feeding. Unlike flatworms, ribbon worms have a complete gut and circulatory system. Above the gut is separate cavity (rhyncocoel) that houses a long proboscis used to capture prey. Some species also have light-sensitive cells at this end on the upper surface. Nemerteans range in length from a millimetre to several metres (although rare intertidally). Nemerteans are difficult to identify to species level by appearance alone, with features used to classify them requiring microscopic examination (e.g. the position of the brain relative to the mouth).

Habitat and ecology
Nemerteans live under rocks, especially where there is an accumulation of sediment, low on the shore and in shallow subtidal areas. Most nemerteans are active hunters, feeding on other worms, crustaceans, molluscs and even small fish. Reproduction is achieved through the shedding of eggs and sperm when two individuals approach each other. The fertilised eggs develop into young either directly or via a larval stage.

Two unidentified nemerteans found in the intertidal zone.

Peanut worms
Phascolosoma spp.
Phylum Sipuncula

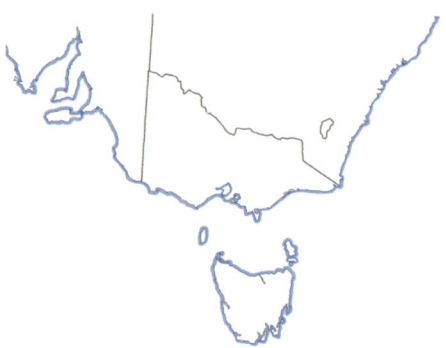

Range
Australia wide

Appearance
Peanut worms are bilaterally symmetrical, unsegmented marine worms. The cylindrical body is made up of a fat trunk section and a thinner extendable section called an introvert that can be withdrawn into the trunk. The introvert has a mouth opening and tentacles at the tip. The trunk may be covered with many tiny wart-like projections (papillae) that give it a rough, leathery appearance. Most individuals found on rocky shores will be less than 50 mm long.

Habitat and ecology
Peanut worms are found in a variety of habitats, such as under boulders, among rock rubble, in crevices and in burrows that have been bored into soft rock. Some species live among colonies of the

Phascolosoma spp.

tubeworm *Galeolaria caespitosa* (p. 74). Peanut worms are either deposit or suspension feeders, depending on the habitat in which they occur. Sexes are separate, with eggs and sperm released externally for fertilisation. The larval stage is free swimming.

Shells
Phylum Mollusca

All molluscs have a head, a muscular foot and a mantle (a layer of tissue that covers the internal organs and secretes the shell). Most also have a shell in some form and a radula (a thin ribbon of rasp-like teeth for feeding that is unique to this phylum).

Chitons (Class Polyplacophora)

Chitons are recognised by their flattened oval appearance, a shell made up of eight overlapping plates (valves) and a leathery surrounding girdle. South-eastern Australian shores are home to a high diversity of chiton species, many of them endemic.

Giant chiton
Plaxiphora albida
Phylum Mollusca, Class Polyplacophora, Family Mopalidae

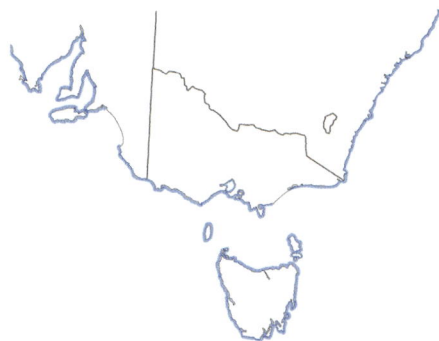

Plaxiphora albida.

Range
Southern WA, SA, Vic., NSW, southern Qld, Tas.

Appearance
P. albida grows up to 13 cm long and has a girdle that is covered with bristles.

Habitat and ecology
This species is common in the lower intertidal zone on shores with moderate to strong wave exposure. Individuals are often found in shallow depressions, where they use a muscular foot to adhere tightly to the rock surface. Homing behaviour, whereby individuals return to the same depression after feeding excursions, has been observed. Chitons use a toothed radula to graze on encrusting algae (pp. 20, 34). Males and females usually release gametes into the water for fertilisation and larval development. Chitons have a trochophore larva that feeds on other plankton until they settle out of the water column to develop into the adult chiton body form.

Ischnochitons
Ischnochiton spp.
Phylum Mollusca, Class Polyplacophora, Family Ischnochitonidae

Range
Ischnochitons are found Australia wide, with more than 30 species recorded from southern Australia. *I. australis* and *I. elongatus* occur throughout south-eastern Australia, whereas *I. versicolor* is less common in SA, *I. lineolatus* is less common in NSW and most records for *I. contractus* are from central SA.

Appearance
Ischnochitons are ovate to elongate and range from 10 to 100 mm in length. Lateral areas of the middle valves are sculptured with raised radial ribs. The tail valve is the longest. The species are difficult to tell apart by colour pattern alone because this can vary within species. Valve sculpturing plus the size, shape and ornamentation of the girdle scales are features that can be used to distinguish species.

Habitat and ecology
Ischnochitons live under boulders and rock rubble in the lower intertidal zone and subtidally. The animals use their muscular foot, as seen in the photograph of a chiton's ventral surface (p. 81), to glide over rock surfaces and move away from light when their rock home is overturned. *I. australis* exhibits a distinctive escape behaviour, sometimes dropping off the disturbed rock and curling into a ball. *I. lineolatus* and *I. elongatus* appear to be more common under fairly smooth rocks on silty reefs.

Individuals are not dispersed evenly in available habitat across a reef, but tend to occur in aggregations under some rocks. Ischnochitons graze on encrusting algae (pp. 20, 34). Sexes are separate and female chitons lay their eggs onto the rock surface for fertilisation and development.

Examples of common *Ischnochiton* species in south-eastern Australia

Ischnochiton australis
I. australis has shiny dark green valves, although these become more eroded and less shiny in larger individuals. This is the largest species of the *Ischnochiton* genus (up to 80 mm long).

Ischnochiton elongatus
I. elongatus is smaller (<35 mm) and more elongate than the other *Ischnochiton* species. The most typical patterning consists of dark valves with a pale central stripe. Girdle scales are medium in size (up to 0.22 mm long).

Ischnochiton versicolor
I. versicolor is usually greenish brown with symmetrical white streaks, but may be uniformly white, fawn or brown. Girdle scales are large (up to 0.48 mm long), and individuals can grow to 65 mm long.

Ischnochiton cariosus
I. cariosus is pale yellow to orange and grows to 50 mm. The anterior valve is strongly nodulated, and the girdle has scales of different sizes that are largest close to the valves and smallest along the outer margin.

Ischnochiton contractus
I. contractus can be any colour, but often has darker markings down the midline and the girdle is usually grey or brown. Length to 40 mm.

(a) *Ischnochiton australis*, (b) *Ischnochiton australis* with an eroded shell, (c) *Ischnochiton elongatus*, (d) *Ischnochiton cariosus*, (e) *Ischnochiton contractus* and (f) *Ischnochiton* sp. ventral view.

Snails (Class Gastropoda)

Gastropod literally means 'stomach foot'; the internal organs are twisted during development so that the stomach sits on top of the muscular foot. Gastropods usually have a coiled shell, although this may be reduced or absent in some groups of this class, which includes common garden snails and slugs.

Abalone, elephant snail and limpets
Abalone
Haliotis spp.
Phylum Mollusca, Class Gastropoda, Family Haliotidae

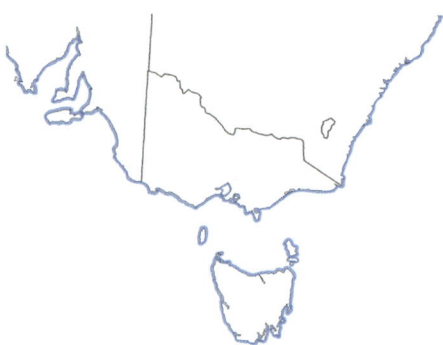

Range
Southern WA, SA, Vic., NSW and Tas.

Appearance
Abalone have a flattened spiral shell that is oval with a row of respiratory holes present near the left margin. A large muscular foot clamps the animal tightly to rock surfaces. Several species occur in southern Australia. Animals of the larger species may grow to 200 mm long, but individuals found intertidally are usually less than 70 mm long.

Habitat and ecology
Most abalone are subtidal, but juvenile blacklip abalone *H. rubra* and individuals of the smaller species *H. scalaris* can be found in the subtidal fringe zone and low intertidal zone, especially on shores within MPAs. Abalone feed on algae,

Haliotis rubra (a) juvenile animal, (b) individuals in a typical crevice habitat and (c) iridescent nacreous layer of the inner shell.

including drift algae, and are eaten by seals. Abalone form the basis of a major commercial fishery in southern Australia, and recreational collecting is limited by fisheries regulations.

Elephant snail
Scutus antipodes
Phylum Mollusca, Class Gastropoda, Family Fissurellidae

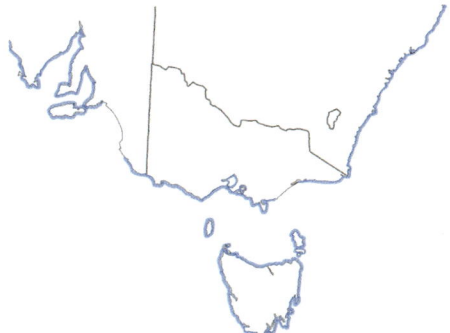

Range
WA, SA, Vic., NSW, Qld, Tas., New Zealand

Appearance
The elephant snail has a large, fleshy, black body with a flat, white shell covering half its upper body surface in the centre of its back. The mantle can completely cover the shell so that, at times, no shell is visible. Two thick tentacles protrude from the anterior end. Individuals can grow up to 100 mm long.

Habitat and ecology
The elephant snail is related to keyhole limpets (p. 84) and is found under rocks, in crevices and in rock pools in the lower intertidal zone and subtidally on shores with moderate wave exposure. This animal feeds on algae and may 'catch' detached bits of algae drifting past. It is more active at night. The flesh of the elephant snail is reported to have an unpleasant taste that would put most people off eating it.

(a) *Scutus antipodes* in typical habitat under a ledge; (b) dorsal and (c) anterior views. (d) *Scutus* shell.

Keyhole limpet

Amblychilepas nigrita
Phylum Mollusca, Class Gastropoda, Family Fissurellidae

Range
WA, SA, Vic., NSW, southern Qld, Tas.

Appearance
Keyhole limpets have a hole in the apex of their shell to allow water containing waste products to exit from the mantle cavity beneath. The shape and size of the opening is a distinguishing feature between species. The shell is smaller than the animal and sits towards its head region. Animals grow to approximately 25 mm. *A. nigrita* has an oblong shell with an oval central hole. The shell is brown with darker radial rays and is sculptured in a beaded radial pattern. The animal is yellow to pale brown (not black, as the Latin name would suggest).

Habitat and ecology
Keyhole limpets avoid sunlight and are found under rocks at low tide level on exposed and semiprotected rocky shores. They are algal grazers. The sexes are separate.

Amblychilepas nigrita.

Rugose slit limpet, cap-shaped false limpet
Montfortula rugosa
Phylum Mollusca, Class Gastropoda, Family Fissurellidae

Montfortula rugosa.

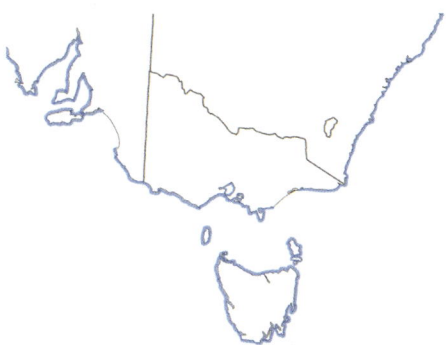

Range
WA, SA, Vic., NSW, Qld, Tas., New Zealand

Appearance
M. rugosa has an oval shell with pronounced radiating ribs. Concentric ridges on the ribs give the shell a granular appearance. The shell apex is central and points towards the posterior end of the animal. The shell is usually off-white but may have a greenish tinge. Not a true limpet, this limpet-like species is related to elephant snails and keyhole limpets (Fissurellidae). Animals can grow up to 20 mm.

Habitat and ecology
M. rugosa is a common limpet of the mid-tide level, often in association with *Galeolaria* (p. 74) or mussels. This species is a generalist grazer, feeding on macroalgae, detritus and microalgae. Individuals can live up to 3 years. Sexes are separate, and eggs and sperm are released into the sea for fertilisation. In New South Wales, the breeding season extends from October until May.

Variegated limpet
Cellana tramoserica
Phylum Mollusca, Class Gastropoda, Family Nacellidae

Range
SA, Vic., NSW, Qld, Tas.

Appearance
This large, very common limpet has an oval-shaped conical shell and grows up to 50 mm long. Colouration can vary considerably, hence the common name. The most usual colouring is orange–brown with occasional ribs of orange or black. As the animals age, the shells erode and host epifauna, such as barnacles and small limpets. Shell shape also varies with wave exposure, with limpets on shores with moderate to high wave exposure having more conical shells.

Habitat and ecology
Having a wide tolerance range to exposure and dampness, this species is found at all tide levels, but is most abundant mid-shore. As a grazer on microscopic algae, it plays a role in controlling algal growth. Many limpets return to the same home 'scar' after feeding excursions. This limpet is a food source for predatory snails, oyster catchers and wrasse. As with all true limpets, gametes (sex cells) are released into the sea for fertilisation. Fertilised eggs hatch into planktonic larvae that swim freely for a few days before settling on a shore.

Cellana tramoserica. The three images illustrate variations in colour pattern and morphology.

Tall-ribbed limpet and pied limpet
Patelloida alticostata, Patelloida latistrigata
Phylum Mollusca, Class Gastropoda, Family Lottiidae

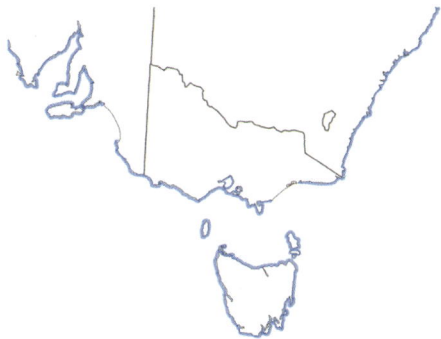

(a) *Patelloida alticostata* and (b) *Patelloida latistrigata*.

Range
P. alticostata: WA, SA, Vic., NSW, Tas.

P. latistrigata: SA, Vic., NSW, Qld, Tas.

Appearance
P. alticostata individuals grow up to 40 mm long. The shell is white and has prominent, radiating ribs with black cross-ribs between them – these are often eroded. The shell margin is strongly scalloped.

P. latistrigata individuals grow to 15 mm long and are a 'long oval' shape, with many individuals narrower at one end. The shell is brown with white rays that generally follow up to 12 radiating ribs (fewer and less defined than in *P. alticostata*). The top of the shell is often eroded.

Habitat and ecology
P. alticostata occurs subtidally and intertidally on shores with moderate to strong wave exposure. It is often found in areas that stay damp during low tide (e.g. crevices, shallow depressions and rock faces receiving wave splash). As with most limpets, this animal feeds by scraping microscopic algae off rocks. The dog whelk (p. 102) and wrasse (fish) feed on this species. Large individuals may be 5 years old. Gametes are released in early summer.

P. latistrigata is common in the mid- to upper tidal zone on most shores, where it occupies small depressions in the rock. It is often closely associated with aggregations of barnacles and/or with air-breathing, pulmonate limpets (*Siphonaria* spp., p. 90). Studies have shown *P. latistrigata* is restricted to these areas by competition for resources with the variegated limpet (p. 86). Breeding occurs between May and November. Large adults are about 3 years old.

Petterd's limpet
Notoacmea petterdi
Phylum Mollusca, Class Gastropoda, Family Lottiidae

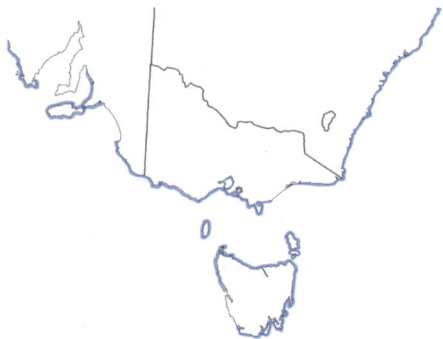

Notoacmea petterdi showing colour variation.

Range
SA, Vic., NSW, southern Qld, Tas.

Appearance
Many of the small limpets (<15 mm long) are difficult to distinguish by sight. *N. petterdi* is dull white or grey with dark radiating bands.

Habitat and ecology
N. petterdi occur on the undersides of overhangs and on vertical rock faces at high tide level on shores with moderate to high wave exposure. Individuals may live for 10 years. In the *Notoacmea* species that have been studied, the sexes are separate, with eggs and sperm being released into the sea during early winter. The fertilised eggs develop into free-swimming larvae that settle after about 8 days. All feed by grazing on microscopic alga. Other species found on rocky shores are *N. alta*, *N. flammea* and *N. mayi*.

Animals of rocky shores | 89

Limpet-like air breathers
Intertidal (ocean beach) slug
Onchidella nigricans
Phylum Mollusca, Class Gastropoda, Family Onchidiidae

Range
Vic., NSW, Tas.

Appearance
O. nigricans is a shell-less, slug-like mollusc that has a tough, leathery 'skin' covering the upper surface and a pair of small tentacles on its head. It grows up to 25 mm long and is dark brown–green mottled with patches of lighter green. The Family Onchidiidae contains air-breathing slugs more closely related to land and freshwater slugs than to other marine sea slugs, which use gills to breathe.

Onchidella nigricans showing colour variation.

Habitat and ecology
The ocean beach slug lives in the mid-tidal zone on shores with moderate wave exposure and is often found in tubeworm encrustations and mussel beds. It is more active during low tides on overcast humid days and at night, sheltering in crevices on sunny days and with the incoming tide. This slug feeds on microscopic algae during low tide.

False limpets

Siphonaria spp.

Phylum Mollusca, Class Gastropoda, Family Siphonariidae

Range

S. diemenensis: WA, SA, Vic., NSW, Qld, Tas.

S. zelandica: WA, SA, Vic., NSW, Qld

S. tasmanica: eastern SA, Vic., southern NSW, Tas.

S. funiculata: SA, Vic., NSW, southern Qld, Tas.

Appearance

False limpets differ from true limpets in that they are air breathers: they have no gills and are related to land snails. However, unlike the ocean beach slug, false limpets can breathe when submerged via folds in the mantle. A slight protrusion of the shell margin on the right-hand side marks the hole through which they breathe (pneumostome).

Several species of *Siphonaria* occur on southern shores. All have shells with radiating ribs. Animals grow up to 25 mm.

(a) *Siphonaria diemenensis*, (b) *Siphonaria funiculata*, (c) *Siphonaroa zelandica* and (d) *Siphonaria tasmanica*.

S. diemenensis has a dark shell with up to 20 pronounced white radiating ribs. The apex is usually eroded.

S. zelandica has a flat white–cream shell with a greenish tinge and a mix of broad and narrow ribs. It is usually found in shallow pools near the high tide level.

S. tasmanica has a conical-shaped bluish shell with alternating concentric dark and light rings and numerous closely spaced ribs. It is usually found on vertical rock faces high on the shore.

S. funiculata has narrow dark brown and white ribs and the shell is more elevated than in the other species. In contrast with the other species, the margin in *S. funiculata* is fairly smooth.

Habitat and ecology
Siphonarian limpets are found across all parts of the shore. Members of the genus *Siphonaria* are hermaphroditic, with internal fertilisation of eggs.

S. diemenensis lays yellow coiled egg masses (see p. 197) in spring and summer from which planktonic larvae hatch. Each egg mass contains thousands of fertilised eggs that hatch into planktonic larvae. The egg masses of *S. zelandica* are similar to those of *S. diemenensis*, whereas *S. tasmanica* and *S. funiculata* produce egg masses that float in the water.

S. zelandica tends to occur on parts of the shore that are frequently inundated with sand. In addition, this species often occurs in association with *S. funiculata*.

S. tasmanica is often found on vertical rock faces. These animals use their radula to feed on macroalgae and return to a home scar after feeding.

Coiled shells
Elongate false ear shell, common ear shell
Stomatella impertusa
Phylum Mollusca, Class Gastropoda, Family Trochidae

Stomatella impertusa.

Range
WA, SA, Vic., NSW, Qld, Tas.

Appearance
The shell of *S. impertusa* is unlike that of other trochid species in that it is elongate and flat. The thin, ovate shell has three whorls, with the apex at the posterior end. Specimens of this species look like young abalone but are distinguished by the lack of respiratory holes in the shell. Shell colour varies (brown, fawn, pink, grey, green, often with yellow triangular markings) and the shell interior is iridescent. Adult animals may grow to 25 mm long.

Habitat and ecology
Individuals of this species tend to be active at night and may be found under rocks in the lower intertidal and shallow subtidal zones. They are often grouped together during the day. They will move to find shade when their rock is turned over. The muscular foot cannot be retracted fully into the shell. The rear part of the foot may be shed as a decoy when an animal is threatened, then regenerated over the next few weeks.

Topshells/winkles
Austrocochlea constricta, ribbed topshell
Austrocochlea porcata, zebra topshell
Diloma concamerata, wavy topshell
Chlorodiloma odontis, checkered topshell
Chlorodiloma adelaidae, Adelaide periwinkle
Phylum Mollusca, Class Gastropoda, Family Trochidae

Appearance
Adult topshells are approximately 15–32 mm tall and 22–28 mm wide at the base. They are taller than or equal to their width at the base and have spiral ridges around the whorls. The operculum that closes over the shell opening (aperture) when the animal is retracted inside the shell is thin, brown and circular.

A. constricta varies from dull grey to dark grey with no bands.

A. porcata has light and dark bands running from the spire to aperture.

D. concamerata has a black shell with pale yellow spots. It is flatter than *A. constricta*, with shell height less than base width, and the ribbing is less pronounced.

C. odontis has a smooth, dark grey shell with chequered yellow spots.

C. adelaidae has a grey to green shell with chequered white to pale yellow spots and narrow spiral ribs.

Habitat and ecology
The ribbed and zebra topshells are found in association with *Bembicium nanum* (p. 97) and *Cellana tramoserica* (p. 86) in the mid-intertidal zone on shores with low to moderate wave exposure. Common and abundant, they graze on the microscopic algae that coat intertidal rock surfaces. The planktonic larvae may drift in the ocean currents for a week before settling.

D. concamerata is found under rocks and boulders on shores with high wave energy.

C. odontis and *C. adelaidae* are found in rock pools, on algae and under rocks in the mid- to lower intertidal zone of shores with moderate to strong wave exposure.

Range
A. constricta: WA, SA, Vic., NSW, Qld, Tas.

A. porcata: SA, Vic., NSW, Qld, Tas.

D. concamerata: south-eastern WA, SA, Vic, Qld, Tas.

C. odontis: SA, Vic., southern NSW, Tas.

C. adelaidae: southern WA, SA, Vic., Tas.

(a) *Austrocochlea constricta* and (b) *A. constricta* aperture. (c) *Austrocochlea porcata*. (d) *Diloma concamerata*. (e) *Chlorodiloma odontis* and (f) *C. odontis* aperture. (g) *Chlorodiloma adelaidae* and (h) *C. adelaidae* aperture.

Warrener, turban shell
Lunella undulata
Previously *Turbo undulatus*

Phylum Mollusca, Class Gastropoda, Family Turbinidae

Range
WA, SA, Vic., NSW, Qld, Tas.

Appearance
This large, round shell resembles a turban and grows up to 50 mm in diameter. The shell is dark blue–green with zigzag white stripes and may be covered with a thin brown layer. The circular aperture is protected by a solid, dome-shaped white cap (operculum).

Habitat and ecology
The warrener occurs in rock pools, crevices and among algae in the lower intertidal zone on shores of any exposure. This snail grazes on algae at lower levels on the shore, and is eaten by birds, fish and people. The small limpet *Hipponyx conicus* is often found on the body whorl near the aperture. The warrener spawns gametes between spring and autumn, with fertilised eggs developing into planktonic larvae.

(a) *Lunella undulata*, (b) *L. undulata* aperture, (c) *L. undulata* in its habitat and (d) *L. undulata* with the limpet *Hipponyx conicus* on its shell.

Pheasant shell, painted lady
Phasianella australis
Phylum Mollusca, Class Gastropoda, Family Phasianellidae

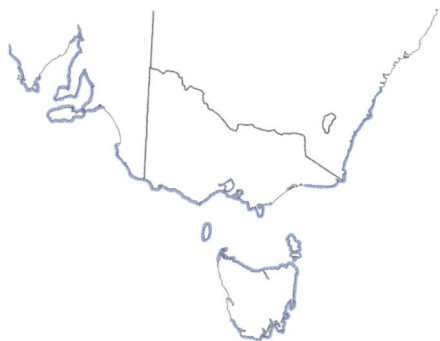

Range
Shark Bay, WA, to Wollongong, NSW, including Tas.

Appearance
This large gastropod grows up to 100 mm long and has a shiny shell covered in a beautiful pattern of red, brown, pink, white and/or purple. A thick, white cap (operculum) blocks the aperture.

Habitat and ecology
The pheasant shell is found in rock pools, rock rubble or the lower tide zone on shores with low to moderate wave exposure. This species is commonly found in association with marine plants and has been observed feeding on sea lettuce (*Ulva* spp., pp. 32–33) and epiphytic algae. Fertilisation of eggs occurs in the water, and planktonic larvae hatch a day later.

Phasianella australis. (a) Dorsal view. (b) Ventral view showing the operculum covering the aperture.

Black crow (Nerite)
Nerita spp.
Phylum Mollusca, Class Gastropoda, Family Neritidae

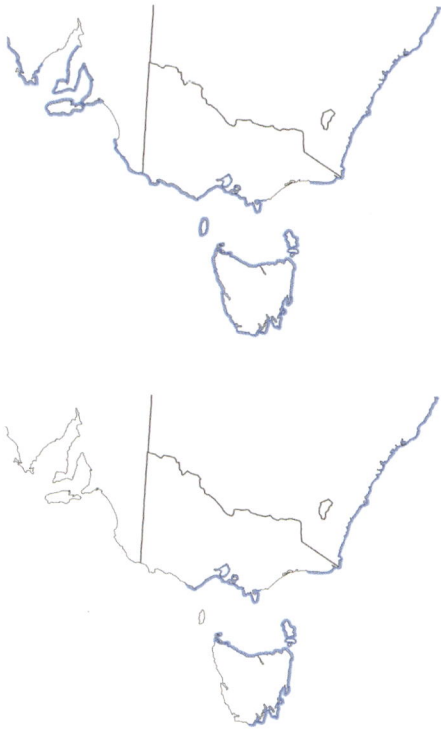

Nerita atramentosa (a) as seen on the shore and (b) ventral surface showing the aperture.

Range
N. atramentosa: Australia wide (top map)

N. melanotragus: central Vic., NSW, southern Qld, Tas. (bottom map)

Appearance
This genus comprises two species of very similar appearance, with some overlap in range. Both have a distinctive globose black shell with a depressed spire and white aperture. The western black nerite (*N. atramentosa*) has a black operculum, whereas the eastern black nerite (*N. melanotragus*) has an orange and black cap (operculum). Both grow to around 30 mm.

Habitat and ecology
The black nerite scrapes microalgae from rocks with its radula and moves and feeds more during night-time low tides. It lays small, flat, white egg capsules from spring to autumn from which planktonic larvae hatch. Adult size is reached at 2 years and the life span is typically 3 years.

Conniwinks

Bembicium nanum,
striped-mouth conniwink

Bembicium melanostoma,
black-mouthed conniwink

Phylum Mollusca, Class Gastropoda, Family Littorinidae

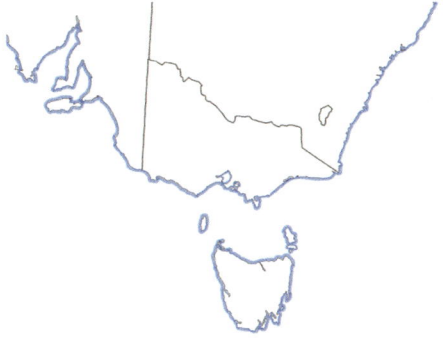

Range
SA, Vic., NSW, Qld, Tas.

Appearance
B. nanum has a small (up to 10 mm in diameter) 'top-shaped' shell, with height less than width, and is cream to yellow with brown stripy markings. The shell of *B. melanostoma* is larger (height to 15 mm) and rougher, with prominent ribs or knobs.

Habitat and ecology
B. nanum is common at high to mid-intertidal zones on shores with moderate to strong wave exposure. *B melanostoma* occupies this niche on sheltered shores and in mangroves.

Both species feed on microscopic algae and are active whenever the rock surface is moist. Sexes are separate and fertilisation is internal. During spring, females spawn soft, kidney-shaped egg capsules, yellow in the case of *B. nanum*, arranged in groups. Each capsule contains more than 100 eggs, from which planktonic larvae hatch.

(a) *Bembicium nanum.* (b) *Bembicium melanostoma.*

Periwinkles
Austrolittorina unifasciata
Afrolittorina praetermissa
Nodilittorina pyramidalis, pyramid periwinkle, nodular periwinkle
Phylum Mollusca, Class Gastropoda, Family Littorinidae

Range
A. unifasciata: WA, SA, Vic., NSW, Qld, Tas.

A. praetermissa: south-eastern WA, SA, Vic., southern NSW, Tas.

N. pyramidalis: northern WA, Mallacoota, Vic., NSW, Qld to Torres Strait

Appearance
A. unifasciata and A. praetermissa grow up to 10 mm long. A. unifasciata is light blue with a dark blue band encircling the whorls of the shell. The whorls of A. praetermissa are covered with dark blue or brown zigzag markings, with the underlying shell colour either light blue or brown. The shape of A. praetermissa is more globose than that of A. unifasciata.

The shell of N. pyramidalis has nodules and a more pointed tip. Length (spire height) up to 20 mm. The shell is light bluish grey with a brown aperture. There are concentric ridges with prominent nodules around the whorls. N. pyramidalis is more elongate than other two species.

Habitat and ecology
All species are abundant at the highest tidal level on all but the most sheltered shores.
N. pyramidalis typically prefers vertical rock faces and occurs at higher elevations than the other members of this family, whereas A. praetermissa can occur lower on the shore. Littorinids survive long periods of emersion by trapping water inside the shell and then secreting a glue-like substance to seal their shells to the rock surface. These snails feed by scraping algae and lichens off the rocks. Eggs in capsules are released into the sea, where they hatch into larvae.

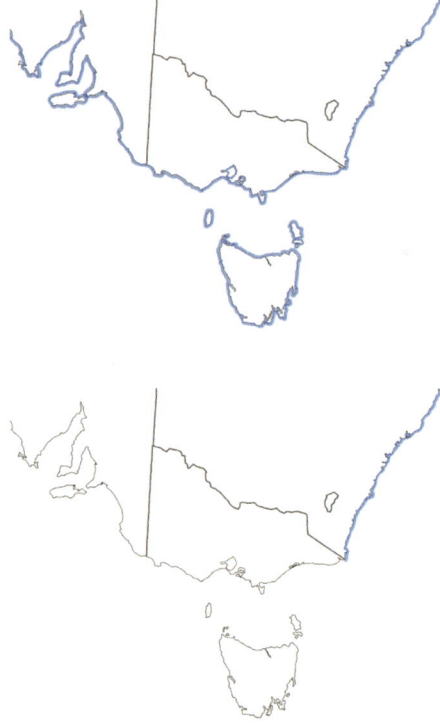

Range for *A. unifasciata* and *A. praetermissa* (top) and for *N. pyramidalis* (bottom).

(a) *Austrolittorina unifasciata* cluster and (b) *A. unifasciata* aperture. (c) *Afrolittorina praetermissa*. (d) *Nodilittorina pyramidalis* and (e) *N. pyramidalis* aperture.

Mitre shells
Isara carbonaria
Phylum Mollusca, Class Gastropoda, Family Mitridae

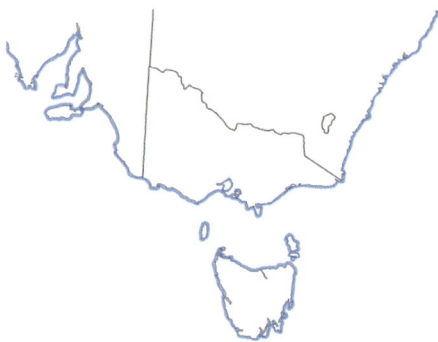

Range
Perth, WA, to Caloundra, Qld, Tas.

Appearance
I. carbonaria has a long, tapered shell with a narrow opening (aperture). Length to 80 mm. The shell is brown to straw coloured, but has a thick dark covering (periostracum) in live animals, with only the tip appearing straw coloured. The animal body is white and has no operculum.

Two other mitre shells within this range are *Isara badia* and *Isara glabra*. *I. glabra* is a similar size (to 90 mm) to *I. carbonaria*, but the shell is thinner and more sculptured, with spiral grooves. The long and tapered shell of *I. glabra* is fawn, with orange–brown in the spiral grooves, under a dark periostracum. *I. badia* is smaller (size to 35 mm), and its shell colour is more reddish amber.

Habitat and ecology
These mitres are found in sand or under rocks in the low-tide zone of mainly exposed rocky shores. They feed on other invertebrates. The preferred prey of *I. carbonaria* and *I. glabra* are peanut worms (p. 78).

(a) *Isara carbonaria*. (b) *Isara badia*.

Veined or purple-mouthed rock shell
Bedeva vinosa
Previously *Lepsiella vinosa*

Phylum Mollusca, Class Gastropoda, Family Muricidae

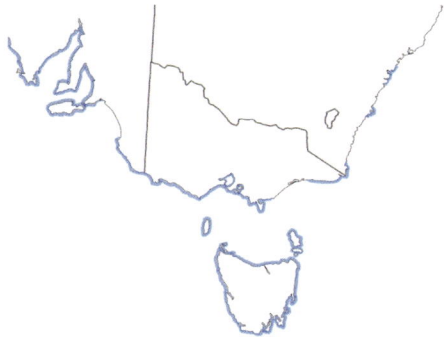

Range
WA, SA, Vic., Tas.

Appearance
The shell shape of this species is variable, but always has a patterning of fine off-white or grey spiral ridges with dark grooves between them. The inside of the shell is a shiny and deep purple (wine coloured), and the cap (operculum) is brown. Individuals may grow up to 20 mm long.

Habitat and ecology
This gastropod occurs at all tide levels, most commonly in association with periwinkles, mussel beds and tubeworm encrustations. *B. vinosa* is a carnivorous snail that uses its radula to drill a small hole in the shell of its prey (other snails, mussels and barnacles) to reach the flesh inside. *B. vinosa* lays small, domed-shaped egg capsules that contain 6–14 fertilised eggs (p. 197). The embryos complete their development within the capsule, and juvenile snails hatch from the capsule.

Bedeva vinosa.

Cartrut shell, dog whelk
Dicathais orbita
Phylum Mollusca, Class Gastropoda, Family Muricidae

Range
WA, SA, Vic., NSW, Qld, Tas.

Appearance
The dog whelk has a heavy shell up to 80 mm long with a short spire. The appearance of the sculpturing 'ruts' on the large body whorl varies from almost smooth to deep, thick spiral ridges. The shell is usually creamy white, although often with a green tinge, and has a dark brown cap (operculum).

Habitat and ecology
D. orbita is abundant in crevices and rock pools low on the shore, or in mussel beds, and is more common on wave-exposed shores. It also occurs subtidally in sheltered bays. This carnivorous snail feeds on other molluscs by drilling a hole in its shelled prey or by forcing the valves of mussels and barnacles apart to reach the flesh inside. *D. orbita* feeds on the turban shell *Lunella undulata* by pushing the operculum aside. *D. orbita* lays many columnar egg capsules joined together in a honeycomb structure. The egg capsules can often have a purple appearance, which is attributable to a possible defensive chemical, known as Tyrian purple. Planktonic larvae hatch from these egg capsules.

Dicathais orbita. (a, b) Colour and form variations. (c) Aperture. (d) *D. orbita* feeding.

Mulberry shell
Tenguella marginalba
Previously *Morula marginalba*

Phylum Mollusca, Class Gastropoda, Family Muricidae

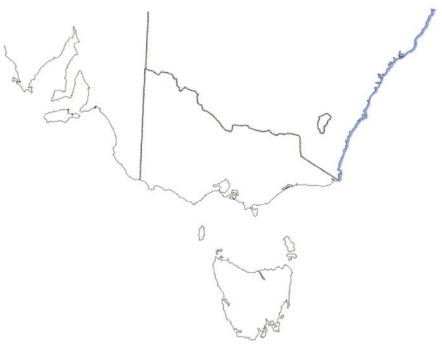

Tenguella marginalba.

Range
NSW, Qld

Appearance
This is a small gastropod with a normal adult size of approximately 20 mm, although it can grow up to 35 mm long. Its robust pale shell has numerous dark, square nodules occurring on spiral ridges. The cream lip of the shell opening is curved and has four teeth on the inside surface.

Habitat and ecology
The mulberry shell is common in crevices and among barnacles on rocky shores in New South Wales. It is also found in oyster beds. This carnivorous snail feeds on barnacles, snails, limpets and oysters by drilling a hole in their shells. Large aggregations of this species often occur among barnacles. The egg masses are hidden (cryptic) and hatch into planktonic larvae.

Lined whelk, chequerboard whelk
Cominella lineolata
Phylum Mollusca, Class Gastropoda, Family Buccinidae

Appearance
This small gastropod grows up to 30 mm long. The shell is light grey, olive green or orange and is encircled by many broken black lines, giving a chequerboard appearance. The images on this page highlight the considerable variation in colour and patterning that occurs within this species.

Habitat and ecology
C. lineolata is common in areas that retain water at low tide, such as rock pools, crevices and algal mats, at any level of the intertidal zone, on shores with moderate to strong wave exposure. A similar species, *C. eburnea*, occurs in sheltered embayments and has a more sculptured shell. *C. lineolata* is a voracious scavenger of both plant and animal matter. It uses a long proboscis for sensing and feeding on prey. During summer, females lay small (5 mm long), white, vase-shaped capsules on short stalks, often laid in in communal masses (p. 198). Each capsule contains 6–14 eggs, from which crawling juveniles emerge.

Range
Geraldton, WA, to Sydney, NSW, Tas.

Cominella lineolata. (a, d) Colour and form variations. (b) Aperture. (c) *C. lineolata* feeding.

Spengler's triton, rock whelk
Cabestana spengleri
Phylum Mollusca, Class Gastropoda, Family Cymatiidae

Cabestana spengleri (a) dorsal and (b) ventral views.

Range
SA, Vic., NSW, southern Qld, Tas.

Appearance
C. spengleri is a large whelk (growing up to 150 mm) with a solid shell that is heavily sculptured with spiral ridges. The ridges (ribs) are fawn, whereas the grooves between them are brown. The outer lip is thickened and has obvious teeth. The operculum is thick and brown. The shell of live animals is covered with a thin yellowish brown outer protective layer called a periostracum.

Habitat and ecology
This whelk occurs in the lowest shore zones (and subtidally to 20 m), usually with cunjevoi (p. 155) or in boulder fields. *C. spengleri* feeds mainly on ascidians, especially cunjevoi. Females lay circular egg masses from November to January that contain long egg capsules from which crawling juveniles emerge.

Tulip shell
Australaria australasia
Phylum Mollusca, Class Gastropoda, Family Fasciolariidae

Range
Southern WA, SA, Vic., NSW, Qld, Tas.

Appearance
This snail has a brown shell and grows to 150 mm. The shell is nodulated with fine spiral ridges. It has a long anterior canal (siphonal groove) and, in live animals, is covered with a thin, brown periostracum. The flesh is a striking dark red.

Habitat and ecology
The tulip shell occurs along the subtidal fringe and subtidally. It is found in crevices during the day, moving onto sandy areas in search of prey during the night. It feeds on a variety of other invertebrates. Egg capsules are laid in clusters under rocks low on the shore in summer. Each capsule is goblet shaped with a frilled edge around the top, and eggs hatch as crawling juveniles (p. 198).

Australaria australasia (a) dorsal and (b) ventral views.

Anemone cone
Conus anemone
Phylum Mollusca, Class Gastropoda, Family Conidae

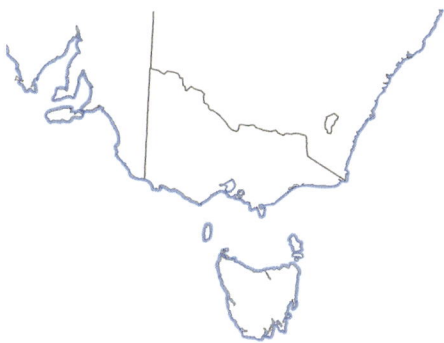

Conus anemone (a) dorsal and (b) ventral views.

Range
WA, SA, Vic., NSW, southern Qld, Tas.

Appearance
This animal has a shell to 80 mm long comprising a large body whorl, a short spire and a long, narrow aperture. The interior of the shell is purplish and the outside has irregular patterns of brown, purple and white. The soft parts of the animal are orange.

Habitat and ecology
Cone shells occur in rock pools and rock rubble from the low-tide zone to depths of 40 m on shores with moderate to low wave exposure. Cone shells have long, hollow radula teeth that look like darts and a deadly toxin that they use to spear and immobilise prey (usually polychaete worms). Females produce a cluster of egg capsules that vary in size and shape (7–19 mm high, 4–12 mm wide) and are laid under rocks in spring. The planktonic larval stage is also bypassed in this species, with well-developed, crawl-away juveniles emerging from the eggs upon hatching. The capsules are flatter, more compressed and lack the distinctive flattened top seen in *Pleuroploca* and *Dicathais* (p. 198).

Live anemone cone shells should not be handled. They are capable of delivering a nasty sting, which could be fatal to someone who is allergic to the toxin. Some tropical species of cone shells have a toxin that is lethal to humans.

Sea slugs and nudibranchs (Class Opisthobranchia)
Sea hares
Aplysia spp.
Phylum Mollusca, Class Gastropoda, Family Aplysiidae

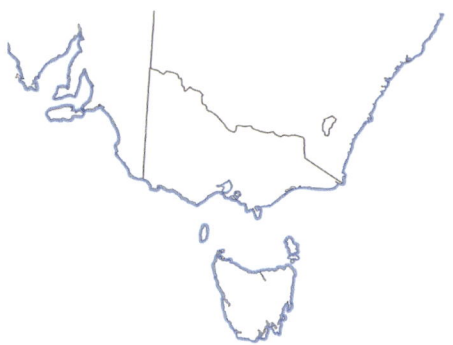

Aplysia sp.

Range
Worldwide; at least five species have been recorded in south-eastern Australia

Appearance
Aplysia is a genus of large sea slugs (to 200 mm) that get their common name from the two long ear-like projections (rhinophores) on the animal's head. Species of this genus typically have a rounded body with frilly parapodia on each side and a distinct head with two oral tentacles and two rhinophores. *Aplysia* species have coloured giant neurons that are relatively easy to work with – they have long been used in neuroscience as a model animal for how neural circuits control behaviours.

Habitat and ecology
Aplysia species are commonly found in shallow waters, tide pools and intertidally on rocky substratum and in seagrass beds. They feed on algae and are eaten by anemones, fish, crabs and rock lobsters. Individuals release a cloud of coloured fluid (often deep purple) when disturbed or threatened. *Aplysia* are hermaphrodites, but rarely self-fertilise. The female sea hares can be abundant in some rock pools, where you may see them munching on seaweed, particularly *Ulva*, in readiness for egg laying. Fertilised eggs are deposited in long strings on rocks or algae, where they hatch free-swimming larvae (p. 198). These metamorphose into a crawling juvenile stage that grows and reaches reproductive maturity at around 12 weeks.

Animals of rocky shores | 109

Short-tailed nudibranch
Ceratosoma brevicaudatum
Phylum Mollusca, Class, Gastopoda, Family Chromodorididae

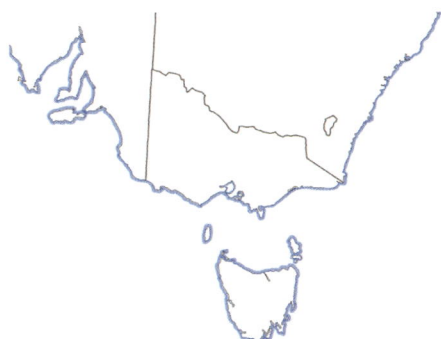

Range
Central east coast of WA to northern NSW, including Tas.

Appearance
Nudibranchs are small, slug-like gastropods without a shell that have feathery external 'gills' and are often brightly coloured. *C. brevicaudatum* has red and purple spots, each with a lighter coloured or white halo, on a yellow, orange, pink or red body. Individuals can grow to 12 cm, but most found intertidally will be smaller. A red glandular lobe is located posterior to the single cluster of dorsal gills. Distasteful chemicals from their prey are stored here, and these act as a defence against predation.

Habitat and ecology
The short-tailed nudibranch occurs in rock pools and the subtidal zone to depths of 120 m. Nudibranchs are carnivores; this species primarily feeds on sponges. As with all nudibranchs, *C. brevicaudatum* species is hermaphroditic. Mating behaviour involves first establishing which of the pair will be male and which will be female. To achieve this, each animal fires its penis at the other; the one that penetrates the body of the other first is the male (i.e. provides the sperm for fertilising the eggs). Eggs are laid in spiral strings and hatch into planktonic larvae.

(a) Side and (b) top views of *Ceratosoma brevicaudatum*. (c) *C. brevicaudatum* with eggs. (Photographs by P. Davis.)

Bivalves (Class Bivalvia)
Common blue mussel
Mytilus spp.
Phylum Mollusca, Class Bivalvia, Family Mytilidae

Mytilus sp. (a) Close-up view. (b) Typical cluster on a wave-sheltered shore.

Range
Temperate regions of Australia and worldwide

Appearance
Most representatives of the genus *Mytilus* in south-eastern Australia are considered to be the introduced species *M. galloprovincialis*, which occurs in temperate waters worldwide. However, recent genetic studies indicate that approximately 20% of *Mytilus* in south-eastern Australia, especially in Tasmania, are the endemic species *M. planulatus*. The species are almost indistinguishable using observable physical characteristics. Adult *Mytilus* grow to 150 mm long and are usually blue–black, but juveniles may be brown. Individual animals are attached to the substratum by thread-like processes called byssus threads.

Habitat and ecology
Blue edible mussels can be found in the lower tidal zone of sheltered shores, often clustered together in mussel beds, and subtidally on pier pilings. They feed by filtering small food particles from the water using their gills. The out-flowing stream of water is used to disperse their numerous planktonic larvae. Mussels may be eaten by predatory fish, snails, sea stars and humans. Pea crabs (*Pinnotheres pisum*) are often found inside their shells. *M. galloprovincialis* is grown commercially for human consumption.

Beaked mussel and flea or horseshoe mussel
Brachidontes rostratus
Xenostrobus pulex
Phylum Mollusca, Class Bivalvia, Family Mytilidae

Range
WA, SA, Vic., southern NSW, Tas.

Appearance
B. rostratus is a large (up to 50 mm long), blue–purple mussel with prominent shell ribs, and often has barnacles and limpets living on its shell.

X. pulex is a smaller (<15 mm long) blue–black mussel with a flattened edge and fairly smooth shell.

Habitat and ecology
Both species form extensive clumps and beds in the mid-intertidal zone on exposed shores, partly due to gregarious settlement of larvae to areas with adults. Beds of *A. rostratus* have many other invertebrates living within them, whereas *X. pulex* beds are flatter and associated with sand. Both species use currents created by their gills to filter feed at high tide. Fertilisation is external and larvae are planktonic.

(a, b) *Xenostrobus pulex* and (c, d) *Brachidontes rostratus*.

Axe head mussel, pygmy mussel, little brown mussel
Xenostrobus securis
Phylum Mollusca, Class Bivalvia, Family Mytilidae

Xenostrobus securis in amongst *Ficopomatus* tubes.

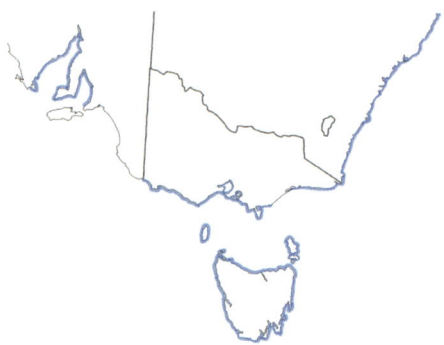

Range
South-west WA to southern Qld, including Tas.

Appearance
The shell of this bivalve is blue–black to brown and elongate with a smooth shiny periostracum. Juvenile animals have characteristic zigzag markings. Adults grow to 30 mm long.

Habitat and ecology
X. securis lives on hard substrata within estuarine/brackish water environments, often forming large colonies, and occurs in and on clumps of *Ficopomatus* (p. 75). It tolerates a wide range of salinity. This Australian/New Zealand species was accidentally introduced into the Mediterranean in the 1990s with aquaculture and has since been found in many other countries, potentially spread through shipping activity. It is extremely invasive.

Sydney rock oyster
Saccostrea glomerata
Phylum Mollusca, Class Bivalvia, Family Ostreidae

Range
Corner Inlet, Vic., eastern Vic., NSW, Qld, Flinders Island and Wynyard, Tas.

Appearance
This large bivalve can grow up to 80 mm long. The lower shell valve is irregular in shape and attaches to the substratum. The top valve is flat. Valves are purple–blue on the outside and pearly white on the inside.

Habitat and ecology
The Sydney rock oyster is often found in the lower intertidal zone on shores with moderate to low exposure to waves, as well as in river estuaries, where it may be commercially harvested. The adult is a filter feeder and may be eaten by predatory gastropods (e.g. *Tenguella marginalba*). Eggs and sperm are released into the sea, where fertilisation occurs. After 2–3 weeks, young oysters (spat) settle into suitable habitats.

(a) *Saccostrea glomerata* aggregation. (b) *S. glomerata* pictured with *Tenguella marginalba*.

Octopus (Class Cephalopoda)
Blue-ringed octopus
Hapalochlaena maculosa
Phylum Mollusca, Class Cephalopoda, Family Octopodidae

The blue-ringed octopus *Hapalochlaena maculosa* (photographs by P. Davis).

Range
WA, SA, Vic., NSW, Qld, Tas.

Appearance
Octopuses have a bulbous head, large eyes, eight arms and a beak within the mouth for feeding. The blue-ringed octopus grows to 100 mm long from top of the mantle (body) to the tip of the arms. Normally this animal is mottled and dull brown, but when disturbed it develops iridescent blue rings on the body and arms.

Habitat and ecology
The blue-ringed octopus is found under loose rocks, in crevices, under seaweed or in dead mussel shells on shores with low to moderate exposure to waves. This animal is active at night. It feeds on crustaceans, shellfish and small fish by grasping its prey with its arms, paralysing it with venom and tearing the flesh with its beak. The venom from the blue-ringed octopus is one of the strongest neurotoxins known and can be fatal to humans because it causes temporary paralysis that stops breathing. However, a blue-ringed octopus will only bite you if provoked. **Do not handle this animal!** The average life span of a blue-ringed octopus is approximately 7 months, with sexual maturity reached at 4 months. Reproduction involves a mating ritual. Females carry the developing fertilised eggs with them in capsules from which crawling or swimming young hatch. There is no planktonic stage.

Maori octopus
Macroctopus maorum
Phylum Mollusca, Class Cephalopoda, Family Octopodidae

Range
SA, Vic., NSW, Tas.

Appearance
M. maorum has a large, oval body shape, large eyes and eight long, thick, muscular arms. The typical colour is dark orange–brown with white spots. The skin has projections that give it a spikey appearance. The mantle length is up to 300 mm, and total length is up to 1 m.

Habitat and ecology
M. maorum occurs on rocky reefs from the intertidal zone to depths over 500 m. This octopus forms lairs, or dens, in rock crevices or in gaps between boulders. Generally active at night, it preys on crustaceans and molluscs, including crabs, rock lobster, abalone and mussels. Females lay numerous small eggs in their lair, which they brood until the eggs hatch into tiny planktonic larvae.

Different views of *Macroctopus maorum*.

Animals with jointed limbs

Phylum Arthropoda, Subphylum Crustacea

Arthropods have a segmented body, jointed limbs and an external skeleton. Most marine arthropods are crustaceans, characterised by having two branches to all limbs except the walking legs, two pairs of antennae on the head and an external skeleton that contains calcium carbonate.

Barnacles *(Class Maxillopoda)*

Barnacles are crustaceans that live fixed to hard surfaces. They have an outside wall made up of calcareous plates. The two dorsal plates open to allow the modified jointed appendages, called cirri, to protrude when the animal is submerged. The cirri are used to create water currents that direct food particles towards the mouth of this stationary animal.

Shore stalked barnacle, hairy stalked barnacle
Ibla quadrivalvis
Phylum Arthropoda, Subphylum Crustacea, Class Maxillopoda, Family Iblidae

Ibla quadrivalvis.

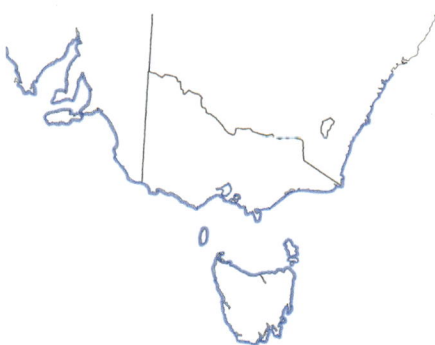

Range
Albany, WA, central SA, Vic., southern NSW, Tas.

Appearance
I. quadrivalvis is a small, stalked (goose) barnacle that grows to 30 mm long. Four plates form a claw-like structure at the end of a yellowish brown stalk that looks as though it is covered in hairs.

Habitat and ecology
This barnacle is found in *Galeolaria* clumps (p. 74) and under rocks at low to mid-tidal levels on sheltered to moderately exposed shores. The free-swimming larval stage depends on yolk reserves from the egg for nutrition. Other stalked barnacles are not covered in hairs, live in the ocean and are often found attached to floating objects, some of which wash ashore (see p. 202).

Surf barnacle
Catomerus polymerus
Phylum Arthropoda, Subphylum Crustacea, Class Maxillopoda, Family Catophragmidae

Catomerus polymerus.

Range
Southern WA, SA, Vic., NSW, Tas.

Appearance
The outside wall of *C. polymerus* is made up of eight large plates surrounded by many smaller plates that decrease in size towards the base, which can be up to 30 mm in diameter.

Habitat and ecology
The surf barnacle is found on shores with strong wave exposure, where it can occur in dense aggregations on vertical rock faces and around mussel beds pounded by surf. Surf barnacles may fall prey to the predatory gastropod *Dicathais orbita*. Barnacles are hermaphrodites, but cross-fertilisation usually occurs between adjacent individuals. Fertilised eggs are brooded until they have developed into planktonic larvae. The larvae of the surf barnacle settle onto the shore during autumn.

Eastern shore (rough) barnacle
Chthamalus antennatus
Phylum Arthropoda, Subphylum Crustacea, Class Maxillopoda, Family Chthamalidae

Range
SA, Vic., NSW, Tas.

Appearance
This barnacle species grows up to 15 mm in basal diameter and has six outer shell plates. The divisions between the six plates are generally clearly visible. The cream or brown shell can appear 'toothed' because the plates are easily eroded.

Habitat and ecology
Rough barnacles are abundant in the upper part of shores with moderate to strong wave exposure. They are often found on the valves of mussels. The animals open their valves to filter food from the water at high tide. Feeding opportunities may be intermittent for individuals in the upper intertidal zone because they may only be submerged a few days each month. Predatory snails feed on barnacles by drilling holes in the shell to eat the animal inside. Larvae are planktonic and settle mainly in summer.

(a) *Chthamalus antennatus*. (b) *C. antennatus* close up showing six shell plates. (c) *C. antennatus* individuals with limbs partially extended.

Honeycomb barnacle
Chamaesipho tasmanica
Phylum Arthropoda, Subphylum Crustacea, Class Maxillopoda, Family Chthamalidae

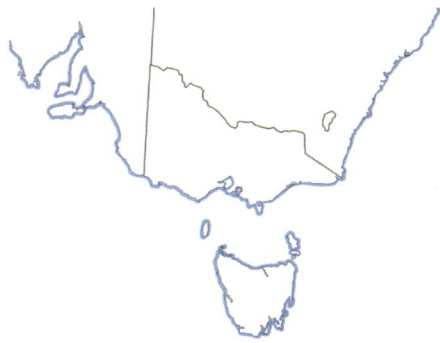

Chamaesipho tasmanica: typical honeycomb appearance (a) and individuals with plates visible (b).

Range
SA, Vic., NSW, Tas.

Appearance
This is a small, greyish barnacle, up to 10 mm high, with a low conical shape and four shell plates. It occurs in such dense clumps that the shell plates of individuals are often indistinguishable, resulting in a honeycomb appearance.

Habitat and ecology
C. tasmanica lives in the mid- to upper intertidal zones on shores with moderate to strong wave exposure. It is found mainly on flat rock surfaces, where individuals can aggregate to densities as high as 90 000 per square metre. Larval settlement on the shore appears to be influenced by larval choice to settle near adults of the same species. Endemic to Australia, this is the most abundant barnacle on southern Australian shores. Two related species, *C. brunnea* and *C. columna*, occur on New Zealand shores.

Rosette barnacle
Tetraclitella purpurascens
Phylum Arthropoda, Subphylum Crustacea, Class Maxillopoda, Family Tetraclitidae

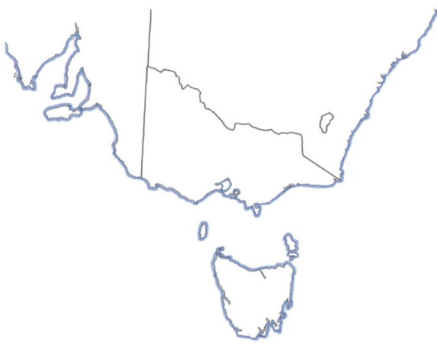

Range
WA, SA, Vic., NSW, Qld, Tas.

Appearance
The shell of this species is flattened, up to 25 mm in diameter around the base and made up of four plates. The aperture is diamond shaped. Shell colour can vary from white to mauve or green.

Habitat and ecology
T. purpurascens occurs with *Chthamalus antennatus* (p. 118) in the mid- and upper tidal zones on shores with moderate to strong wave exposure. It lives mainly in crevices and underhangs in shaded areas. It is prone to 'drying out', and is also found on pier pilings. Little is known about its ecology.

Tetraclitella purpurascens (a) typical cluster and (b, c) two variations in outer plate morphology.

Rosy barnacle, common surf barnacle
Tesseropora rosea
Phylum Arthropoda, Subphylum Crustacea, Class Maxillopoda, Family Tetraclitidae

Tesseropora rosea.

Range
Point Lonsdale, Vic., NSW, southern Qld, north-east Tas.

Appearance
The common surf barnacle has a shell made up of four distinct plates. It grows up to 30 mm in diameter around the base and 15 mm high. Older individuals have worn shells with a distinct pink tinge ('rosea'); younger individuals are often grey–white with a pink tinge at the top of the plates.

Habitat and ecology
T. rosea occurs in the lower reaches of the mid-tide zone on shores with strong wave exposure. It is conspicuous on sloping shores in New South Wales, but rare in Victoria. In flat rocky areas, it often occurs with *Catomerus polymerus* (p. 117). The common surf barnacle is commonly eaten by predatory snails. It has a clearly defined breeding season in summer and early autumn. Larvae settle on the shore from summer through to winter.

Estuarine barnacle, modest four-plated barnacle
Austrominius modestus
Previously *Elminius modestus*

Phylum Arthropoda, Subphylum Crustacea, Class Maxillopoda, Family Archaeobalanidae

Austrominius modestus.

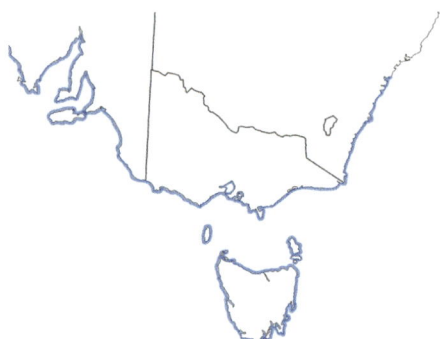

Range
South-west WA to Sydney, NSW, Tas.

Appearance
This small barnacle is pale grey and has four distinct ridged basal plates. It grows to 15 mm in diameter around the base.

Habitat and ecology
A. modestus occurs in the upper to mid-shore zones on sheltered shores (estuaries and sheltered embayments), cemented firmly to rocks, wood and other hard substratum, including the pneumatophores of mangroves. This species appears to be tolerant of turbid, estuarine water (i.e. a wide range of salinity). As suspension feeders, these barnacles use rhythmic beating of their legs to draw water and associated organisms and organic matter into the shell. Individuals grow rapidly and are reproductive in their first year. Eggs hatch into free-swimming nauplii larvae.

Native to New Zealand, and possibly Australia, *A. modestus* has become a pest species in the Northern Hemisphere. It spread to Britain and Europe on the hulls of ships or in bilge water, possibly during the Second World War.

Animals of rocky shores

Slaters, shrimp and crabs (Class Malacostraca)

The basic body plan of this group is a head of six fused segments, a thorax of eight segments and an abdomen with six or seven segments and a tail. Most segments possess a pair of jointed appendages. This class includes most of the crustaceans typically encountered on rocky shores.

Sea slaters (Order Isopoda)

Marine pill bug
Zuzara sp.
Phylum Arthropoda, Subphylum Crustacea, Class Malacostraca, Family Sphaeromatidae

Zuzara sp. (a) dorsal view and (b) rolled into a ball.

Range
SA, Vic., NSW, Tas.

Appearance
Pill bugs, slaters and lice are isopods, crustaceans that have a flat appearance when viewed from the side and do not have a carapace (which lobsters, shrimp and crabs do). Unlike in many other crustacean groups, the eyes in sea slaters are not on stalks. Sea slaters are closely related to the slaters that occur in gardens under pot plants, wood, rocks and/or other objects. The tails of marine species can have various shapes (morphologies), ranging from tails that look like elaborate fans to those that have very sharp, needle-like spines. Many sea slaters can roll themselves into a ball if threatened.

The posterior appendages (uropods) of *Zuzara* are quite large relative to the rest of the body. Individuals are usually around 10 mm long, but may grow to 20 mm.

Habitat and ecology
Found under rocks in the intertidal and shallow subtidal zones, often in groups. Found in various habitats, including soft sediment, seagrass, algal beds and reefs. Some species congregate in large numbers under rocks in the intertidal and shallow subtidal zone. They are mostly scavengers, feeding on decaying algal fragments.

Australian shore slater

Ligia australiensis

Phylum Arthropoda, Subphylum Crustacea, Class Malacostraca, Family Ligiidae

Range
SA, Vic., central NSW, Tas.

Appearance
The body of this shore slater is up to 30 mm long, elongate, flattened and segmented with a pair of appendages on each segment. The eyes and antennae are prominent.

Habitat and ecology
In Victoria, the shore slater is common among stones and broken boulders above the intertidal zone on shores with any wave exposure. In other states, the shore slater usually occurs in sheltered areas or estuaries. Its flattened body shape allows it to move easily under stones and into cracks and crevices. This is a particularly quick isopod that will disappear quickly under rocks when disturbed. The shore slater is related to fish lice, but is a scavenger and harmless to humans. As

Ligia australiensis.

with most isopods, sexes are separate. Mating occurs during moulting, and females brood the fertilised eggs, which hatch as larvae that resemble the adult.

Animals of rocky shores | 125

Sea centipede, kelp lice
Paridotea ungulata
Euidotea bakeri
Phylum Arthropoda, Subphylum Crustacea, Class Malacostraca, Family Iodeteidae

Range
P. ungulata: gulfs of SA to southern NSW and around Tas.

E. bakeri: south-west WA to southern NSW and around Tas.

Appearance
Sea centipedes resemble terrestrial centipedes but are not as long and do not move with snake-like movements. They have a long, thin body that changes colour to match the plants the sea centipedes are living on, usually green or brown, for camouflage. Sea centipedes have seven pairs of legs with sharp toes that can get a solid grip. *P. ungulata* has a concave notch at the end of the final posterior segment, whereas the tail of *E. bakeri* is pointed. *P. ungulata* can grow to 40 mm long, whereas *E. bakeri* is usually about half this size.

Habitat and ecology
These two species can co-occur in seagrass and algal beds in sheltered bays and on the exposed coast. They are herbivores.

(a) *Paridotea ungulata*. (b) Dorsal and (c) ventral views of *Euidotea bakeri*.

Shrimp and crabs (Order Decapoda)
Glass shrimp, estuary shrimp, rock pool shrimp
Palaemon dolospinus, Palaemon serenus
Phylum Arthropoda, Class Malacostraca, Family Palaemonidae

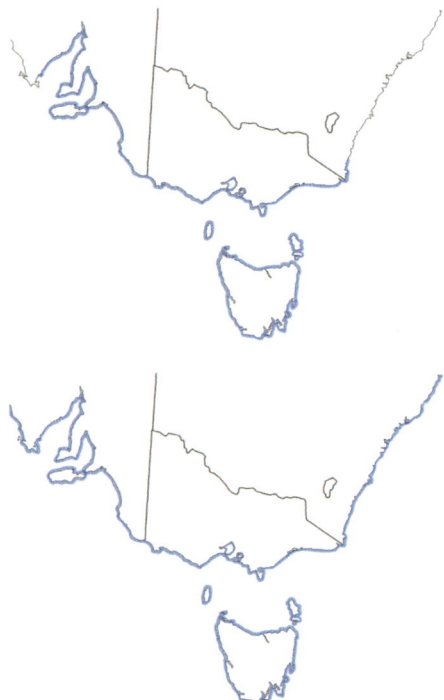

(a) *Palaemon dolospinus*. (b) *Palaemon serenus*. (c) *P. serenus* in habitat (photograph c by P. Davis).

Range
P. dolospinus: SA, southern NSW, Vic., Tas. (top)

P. serenus: south-west WA, SA, Vic., NSW, southern Qld, Tas. (bottom)

Appearance
Members of this family are transparent, with variable colouration and spines on the carapace just below the eye. The estuary shrimp (*P. dolospinus*) is characterised by thin, dark lines that wrap around the abdomen, and is more common in sheltered bays and estuaries. The rock pool shrimp (*P. serenus*) tends to be more common on exposed coasts and has a characteristic red collar around its larger pair of chelipeds.

Habitat and ecology
These shrimp can be locally abundant and a conspicuous component of the invertebrate fauna among seagrass and mixed algal beds. They use their chelipeds to feed on microscopic algae and organic matter, but will also scavenge on carrion. Mating involves the transfer of sperm between males and females, and fertilised eggs are brooded by the female under her abdomen. The eggs develop into planktonic larvae.

Common hermit crab
Paguristes frontalis
Phylum Arthropoda, Subphylum Crustacea, Class Malacostraca, Family Diogenidae

Paguristes frontalis.

Range
Southern WA, SA, Vic. to Wilsons Promontory

Appearance
Hermit crabs differ from other decapod crabs in that only the second and third pairs of legs are used for walking; the fourth and fifth pairs are modified for specific functions. In addition, the abdomen is relatively large, soft and coiled.

P. frontalis is larger than most hermit crabs found in south-eastern Australia, growing to 80 mm. Individuals from South Australia and Victoria have a red body and cream claw tips, whereas those from Western Australia are purple–brown.

Habitat and ecology
Hermit crabs inhabit empty gastropod shells, moving into progressively larger shells as they grow. They retreat into their shell when disturbed. *P. frontalis* is found on reefs from the low intertidal zone to around a depth of 15 m depth. These crabs are omnivores and more active at night.

Spiny porcelain crab, false crab
Petrocheles australiensis
Phylum Arthropoda, Subphylum Crustacea, Class Malacostraca, Family Porcellanidae

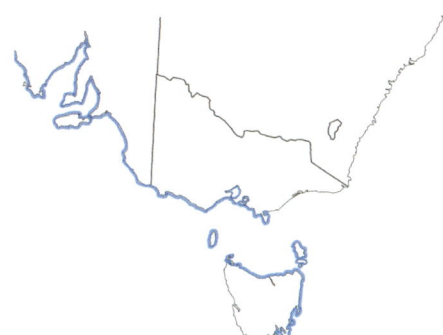

Petrocheles australiensis.

Range
SA, Vic., Tas.

Appearance
This is a squat, pear-shaped crab with very long pincers (chelipeds) that is more related to hermit crabs than true crabs. The fifth pair of walking legs is reduced and tucked away along the carapace. The large, flattened chelipeds are spiny, unlike in other members of this family. Individuals can grow to a carapace length of 23 mm.

Habitat and ecology
A cryptic species that occurs in subtidal rocky reef habitat, in amongst crevices or under rocks. The large claws are used for defence, rather than feeding. These crabs are probably filter feeders because the feeding limbs are covered in numerous long hairs (setae).

Hairy stone crab
Lomis hirta
Phylum Arthropoda, Subphylum Crustacea, Class Malacostraca, Family Lomisidae

Range
SA, Vic. to Wilsons Promontory, southern NSW, Tas.

Appearance
L. hirta is the only species in the Family Lomisidae. Individuals of this species have a triangular carapace up to 20 mm across. The top (dorsal) surface is rough, hairy and usually grey–brown. The distinctive antennae are bright blue. *L. hirta* has three pairs of walking legs in addition to its pincers. The fifth pair of legs is small and hidden from sight. True crabs (Brachyura) have pincers, then four pairs of walking legs that are similar in appearance to each other.

Lomis hirta (a) dorsal and (b) ventral views.

Habitat and ecology
The hairy stone crab lives under rocks at lower tidal levels on shores with moderate wave exposure. It feeds by filtering planktonic plants and animals from the water. Females carry eggs under the abdominal flap.

Ramshorn crab
Naxia aries
Phylum Arthropoda, Subphylum Crustacea, Class Malacostraca, Family Majidae

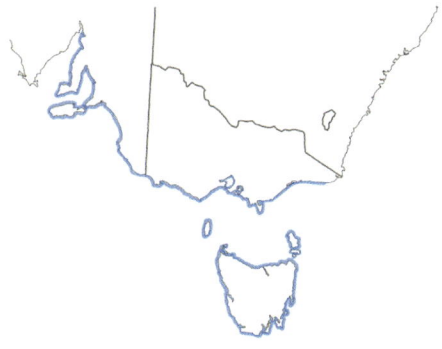

Range
SA, Vic., Tas.

Appearance
The carapace (up to 37 mm wide) has large spines and stiff curved hairs (like velcro) that are used to attach seagrass, algae and other marine debris. The various species within this genus can be distinguished by the two spines on the rostrum and the spines directly above the eye. The carapace is teardrop/pear shaped. The algae and other items attached to the carapace serve as camouflage, but some species of algae may deter predators and some may be harvested as food.

Habitat and ecology
The ramshorn crab can be found under rocks on intertidal reef, seagrass and algal beds in sheltered bays. It feeds on algae and detritus.

Naxia aries (a) dorsal and (b) ventral views.

Bear seaweed crab, hairy seaweed crab
Notomithrax ursus
Phylum Arthropoda, Subphylum Crustacea, Class Malacostraca, Family, Majidae

Notomithrax ursus (photograph by G. Quinn).

Range
SA to Newcastle, NSW, including Tas.

Appearance
A very bristly crab with numerous setae (hairs) on the carapace and legs. The carapace is longer than wide and can be up to 40 mm wide. This crab's ability to remain still, together with its thick cover of decorative material, often makes it very difficult to find. Camouflage is achieved by using fresh seaweed picked from the pool in which the crab is sheltering. The crab coats the cut end of the seaweed with a salivary 'glue', then uses its long legs to place the prepared seaweed on its spiny dorsal surface.

Habitat and ecology
This crab is found under rocks and among seaweed on moderately exposed and exposed coastlines, as well as on shallow subtidal reef. *N. ursus* is an omnivore, feeding on seaweeds, small shellfish and detrital material. This species of seaweed crab is endemic to south-east Australia and New Zealand.

Estuarine sea spider
Amarinus laevis
Phylum Arthropoda, Subphylum Crustacea, Class Malacostraca, Family Hymenosomatidae

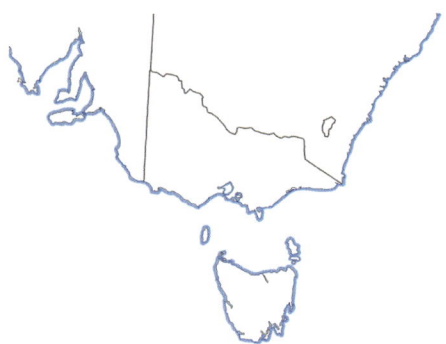

***Amarinus laevis* male** (photograph by D. Squire).

spider-like legs, hence its common name. In some estuaries, the males have a conspicuous orange to red soft swelling at the junction of the pincers. The pincers may look ominous, but are harmless.

Range
Southern WA, to southern Qld, including Tas.

Appearance
A small brown-to-grey crab with a carapace up to 22 mm wide (up to the size of a 10-cent piece). This species has a flat, round carapace and

Habitat and ecology
A. laevis is confined to estuarine environments and inlets, in amongst and under rocks, tubeworm colonies and in wood and rock crevices. In some regions, the animals are sold commercially for bait. They are selective deposit feeders, but will also scavenge on carrion.

Three-pronged flat-back crab, false spider crabs, flat-backed crabs
Halicarcinus ovatus
Phylum Arthropoda, Subphylum Crustacea, Class Malacostraca, Family Hymenosomatidae

Range
From central WA to southern Qld, including Tas.

Appearance
H. ovatus is a small crab with a carapace up to 10 mm wide, typically no larger than thumb nail. It has a flat, round carapace that is often mottled with flecks to blend in with the surrounding habitat. The spider-like legs are more delicate in appearance than those of the estuarine spider crab above. The various hymenosomatid crabs can be distinguished by looking at the size of their 'nose' (rostrum). *H. ovatus* has a short, pointy rostrum and three spikes between the eyes.

Habitat and ecology
This species lives among rock rubble and rocks within seagrass and algal beds in sheltered and exposed coasts. It is a selective deposit feeder, using its chelipeds to pick up detritus.

Halicarcinus ovatus (a) dorsal and (b) ventral views.

Two-spined crab
Litocheira bispinosa
Phylum Arthropoda, Subphylum Crustacea, Class Malacostraca, Family Litocheiridae

Range
Albany, WA, SA, Port Phillip and Western Port bays, Vic., Sydney, NSW, Tas.

Appearance
This small crab has a smooth, shiny carapace of mottled brown colouration that is up to 15 mm wide. The lateral edges of the carapace are straight with a single spine below the eye, a feature that can be used to distinguish it from shore crabs (pp. 138–141). There is also a sharp spine on the large claw.

In most crabs the abdomen is reduced to a flap that is folded under the crab's body (thorax). The sex of a crab can be determined by the width of this flap: wide in females, narrow in males. Females brood fertilised eggs under this flap.

Habitat and ecology
L. bispinosa is found under rocks on sheltered muddy shores. It feeds on detritus. This species is native to Australia.

(a) *Litocheira bispinosa* dorsal view. (b) Ventral view of a female *L. bispinosa*.

Common (European) shore crab
Carcinus maenas
Phylum Arthropoda, Subphylum Crustacea, Class Malacostraca, Family Carcinidae

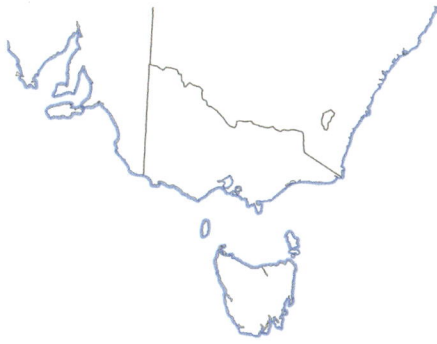

Range
SA, Vic., southern NSW, Tas., worldwide.

Appearance
The carapace of *C. maenas* is deeply notched on the front and sides, and is usually a rich green or brown colour. The carapace measures up to 50 mm across at its widest point near the front and tapers sharply to be narrower at the back.

Habitat and ecology
C. maenas is found under rocks in the upper intertidal zone on both exposed and sheltered shores. A wide salinity tolerance allows it to thrive in estuaries. It is primarily a carnivore that eats small molluscs and other crabs. *C. maenas* is thought to have been introduced to Australia from Europe, possibly as larvae in ballast water. Listed among the top 100 'world's worst invasive alien species', its impact on native fauna is unknown, but thought to be significant. *Nectocarcinus integrifrons* (p. 134) is a related native mudflat species.

(a) *Carcinus maenas*. (b) Male *C. maenas*. (c) *C. maenas* mating.

Swimmer crabs
Nectocarcinus integrifrons,
rough rock crab, seagrass swimmer crab
Nectocarcinus tuberculosus,
red swimmer crab, velvet crab
Phylum Arthropoda, Subphylum Crustacea, Class Malacostraca, Family Carcinidae

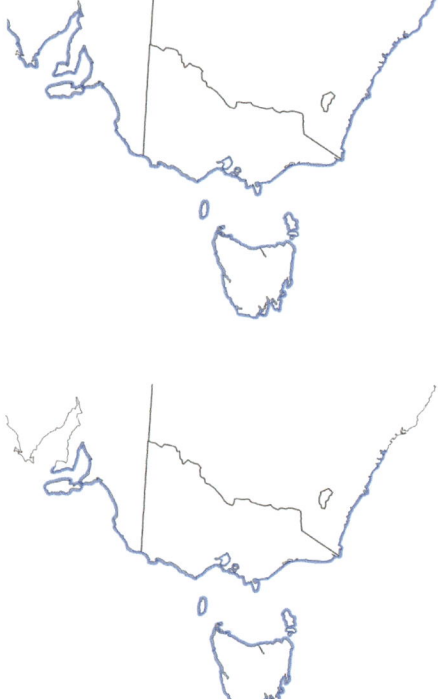

Range for *N. integrifons* (top) and *N. tuberculosus* (bottom).

Range
SA, Vic., NSW, Tas.

Appearance
N. integrifrons has distinctive black fingers (claws) and tends to have a two-toned carapace, which consists of an anterior band of red-to-purple and a posterior band of yellow/cream with purple/red specks. This species can reach of carapace width of up to 80 mm. The front margin of the carapace between the eyes is smooth. This crab can be quite aggressive.

N. tuberculosus has similar colouration to *N. integrifrons* but lacks the distinct two-toned carapace; *N. tuberculosus* also has a small notch between the eyes. *N. tuberculosus* grows to a carapace width of around 90 mm. The common name, velvet crab, comes from the cover of soft, fine hair on the carapace.

Both species have flattened hind legs that help with swimming.

Habitat and ecology
N. integrifrons tends to occur on shores sheltered from waves, in amongst rock rubble or seagrass beds, and can reach high numbers at some locations. Recently moulted and/or dead crabs or the carapace are often washed ashore in Port Phillip Bay. In contrast, *N. tuberculosus* occurs on more exposed coastlines that experience higher wave exposure, and is more common on rocky reef habitat. There is mixed information regarding the diet of *N. integrifrons*, with reports of carnivory and herbivory, but it is likely that both species are also opportunistic scavengers. During breeding, the male crab carries the female on his ventral side until they have finished mating, which occurs after the female has moulted.

(a) *Nectocarcinus integrifrons* in habitat. (b) Ventral view of *N. integrifrons* male. (c) *N. tuberculosus* in habitat. (d) Dorsal and (e) ventral views of female *N. tuberculosus*.

Black-finger crab, iron crab, eastern reef crab
Ozius truncatus
Phylum Arthropoda, Subphylum Crustacea, Class Malacostraca, Family Eriphiidae

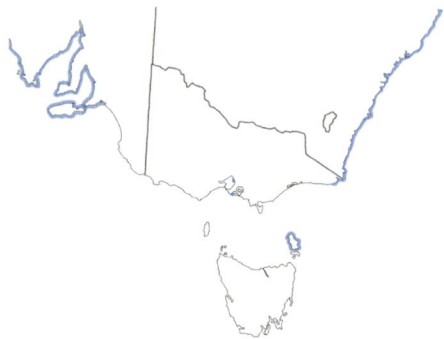

Range
SA, central and eastern Vic., NSW, southern Qld, Flinders Island

Appearance
A robust, slow-moving, dark brown crab with large claws that have dark brown tips (or fingers). Four broad teeth occur along the edge of the carapace. Individuals of this species tend to curl their legs under their body when disturbed (play dead). Females carry eggs under their tail (abdomen; see photograph).

(a) Ventral view of a female *Ozius truncatus* carrying eggs under her abdomen. (b) Dorsal view of *O. truncatus*.

Habitat and ecology
This species can be very common under rocks in soft-sediment habitat in Port Phillip Bay. It is a predatory crab that feeds on snails and other crabs.

Swift-footed crab
Leptograpsus variegatus
Phylum Arthropoda, Subphylum Crustacea, Class Malacostraca, Family Grapsidae

Leptograpsus variegatus (photograph by R. Koss).

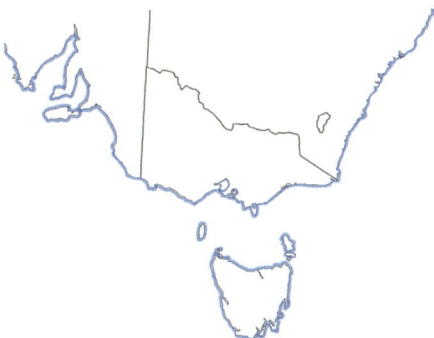

Range
WA, SA, Vic., NSW, Qld, Tas.

Appearance
This crab has a steel-grey carapace up to 80 mm across, with a series of parallel ridges running across the shell. The chelipeds are purple.

Habitat and ecology
L. variegatus lives on shores with strong wave exposure at and above the high-tide level in crevices or scuttling over rocks. It feeds on plant matter scraped from rock surfaces, but also eats mussels and barnacles. L. variegatus is capable of rapid, agile movements and can be aggressive, although it is mainly active at night. Females carry eggs under their abdominal flap during November–February. A similar species, the burrowing shore crab Leptograpsodes octodentatus, is smaller and occurs near the high-tide line on exposed and sheltered shores. It has one to three notches on each side of its dark, rough-textured carapace and is far less agile.

Shore crabs

Cyclograpsus audouinii,
smooth shore crab

Cyclograpsus granulosus,
purple shore crab

Phylum Arthropoda, Subphylum Crustacea, Class Malacostraca, Family Varunidae

(a) *Cyclograpsus audounii.* (b) *C. audounii* ventral view. (c) *Cyclograpsus granulosus.*

Range
C. audouinii: WA, SA, Vic., NSW, southern Qld, Tas. (top map)

C. granulosus: SA, Vic., southern NSW, Tas. (bottom map)

Appearance
The carapace is smooth, up to 35 mm across and roughly square in both species. *C. audouinii* can be distinguished by tufts of hairs between the legs and is speckled brown to grey, whereas *C. granulosus* is more often a mottled purple.

Habitat and ecology
Both species are abundant under rocks on shores with moderate wave exposure. *C. audouinii* lives in the upper intertidal zone and *C. granulosus* lives in the mid-intertidal zones. Both species are scavengers and are themselves eaten by fish and birds. A planktonic larval stage lasts 30–45 days. These two species interbreed to form hybrids.

Notched shore crab
Paragrapsus quadridentatus
Phylum Arthropoda, Subphylum Crustacea, Class Malacostraca, Family Varunidae

Paragrapsus quadridentatus (a) dorsal and (b) anterior views.

Range
Adelaide, SA, Vic., Tas., Sydney, NSW

Appearance
The carapace of this species is roughly square to a width of 30 mm and is a speckled sandy brown with one distinct notch on each side.

Habitat and ecology
P. quadridentatus is often found on sand under rocks in the mid- to lower tide zones on shores with moderate to high wave exposure and is extremely common on boulder shores. This species is a scavenger and is, in turn, eaten by birds and fish. It is active at high tide and can be very aggressive if disturbed at low tide. Two other species of *Paragrapsus* are found on sheltered, muddy shores: *P. gaimardii* and *P. laevis* (p. 191). These species have larger, dark brown carapaces with two distinct notches on each side.

Little shore crab
Brachynotus spinosus
Phylum Arthropoda, Subphylum Crustacea, Class Malacostraca, Family Varunidae

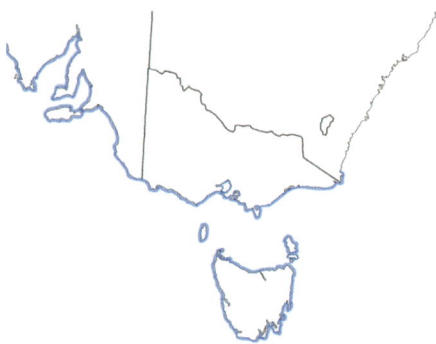

Range
SA, Vic., Bermagui, NSW, Tas.

Appearance
This crab is quite small compared with other grapsid crabs, with a carapace up to 17 mm across (width) and roughly square. There are typically three small, sharp notches on each side of the carapace. The roughened surface varies in colour from green to black, often with two white spots on each side.

(a) *Brachynotus spinosus* in a mussel bed. (b) *B. spinosus* dorsal view.

Habitat and ecology
The little shore crab may be found under rocks or near rock pools at any level on shores with low to moderate wave exposure. This crab is a scavenger, eating a wide range of food. It is preyed on by birds and fish, and is active mainly when covered by high tide.

Asian shore crab, Japanese shore crab
Hemigrapsus sanguineus
Phylum Arthropoda, Class Malacostraca, Family Varunidae

Hemigrapsus sanguineus (a) dorsal and (b) anterior views.

Range
Port Phillip Bay, Victoria

Appearance
This recently discovered invasive species could easily be overlooked as a native shore crab, particularly the juveniles. The adults have a large, square carapace, characterised by three large spines on each side, that can reach 30 mm in width. It has large, mottled claws with purple spots and a distinct bulb of soft tissue between the claws (pulvinus). The legs typically have characteristic black banding (stripes).

Habitat and ecology
This species is currently known from a small number of rock reefs inside Port Phillip Bay, and probably arrived via ship ballast waters. It has a native range in the north-west Pacific, but it has also spread to several other parts of the world, including Europe and North America. Research into the impacts of this species on native species in Port Phillip Bay has only recently commenced, but, given their large claws and carapace, these crabs are likely to be having an impact on local marine ecosystems. This species feeds on a range of smaller invertebrates. Females can be highly fecund (i.e. carry tens of thousands of eggs) and typically start producing eggs in spring. The reproductive behaviour in Port Phillip Bay remains unknown. The eggs hatch into planktonic larvae.

Red bait crab, red rock crab
Guinusia chabrus
Previously *Plagusia chabrus*

Phylum Arthropoda, Subphylum Crustacea, Class Malacostraca, Family Plagusiidae

(a) *Guinusia chabrus* frontal view. (b) Dorsal view of a juvenile. (Photograph a by P. Davis.)

Range
WA, SA, Vic., NSW, southern Qld, Tas.

Appearance
The carapace of this crab is bright orange to red and grows up to 70 mm across. Distinctive clefts are visible between and to each side of the eyes, and the legs bear numerous sharp spines.

Habitat and ecology
G. chabrus is found in crevices and rock pools in the lower intertidal zone and subtidally on shores with moderate to strong wave exposure. The crab is active at high tide, feeding on algae, sponges, bryozoans and molluscs. A dense covering of short hairs over the carapace helps this crab escape from one of its predators, the octopus. This crab is also the prey of birds, fish and humans. Breeding occurs in spring and summer, and eggs are brooded under the abdominal flap.

Lace coral
Phylum Bryozoa

Bryozoans are colonies of many tiny individuals (<1 mm), called zooids, that are generally attached to hard surfaces, like rocks. A zooid looks like a tiny U-shaped worm with a crown of tentacles, called a lophophore. Each individual lives in its own 'shell', but are interconnected for the exchange of nutrients and other chemicals. Different types of zooids work together to support the colony. Some use the lophophore to filter food from the surrounding water, providing nutrition to the rest of the colony, whereas others have reproductive or defensive roles.

Red bryozoan
Mucropetraliella ellerii
Phylum Bryozoa, Class Gymnolaemata, Family Petraliellidae

Range
SA, to central NSW, Tas.

Appearance
M. ellerii is a vivid red, roughly textured, calcareous, encrusting colony that forms approximately 1-mm thick sheets up to 50 mm in diameter. The colonies feel rough, like a fine rasp.

Habitat and ecology
M. ellerii is restricted to the moist, shaded undersides of large rocks or beneath ledges in the lower intertidal zone on shores of all types of wave exposure. Each microscopic individual in a bryozoan colony has a crown of tentacles (lophophore) that can be used to filter food from

Mucropetraliella ellerii.

the surrounding water. Many individuals in a colony brood planktonic larvae in external chambers. Colonies are often overgrown by algae, sponges and ascidians.

Animals with tube feet
Phylum Echinodermata

Phylum Echinodermata includes sea stars, sea urchins, brittle stars, feather stars and sea cucumbers. All except sea cucumbers have a spiny skeleton. A unique feature of this phylum is a hydraulic water vascular system that circulates fluid around the body for gas exchange and nutrition. Connected to this system are fluid-filled tube feet that are used in locomotion.

Urchins (Class Echinoidea)
Purple sea urchin
Heliocidaris erythrogramma
Phylum Echinodermata, Class Echinoidea, Family Echinometridae

Range
WA, SA, Vic., NSW, Qld, Tas.

Appearance
This urchin has a circular and slightly flattened shell, called a test, up to 100 mm across, that is covered in sharp, dark-coloured spines, each up to 30 mm long. The colour of the test can be pink, dark red, light purple, green or cream.

Habitat and ecology
H. erythrogramma is found in the lower intertidal zone and subtidally on shores of moderate to strong wave exposure. It lives in rock pools, under ledges or in holes in the rock and is common subtidally among kelp and seagrasses in sheltered bays. Males and females release gametes into the sea during summer, with fertilisation resulting in planktonic larvae. This species feeds on encrusting and drift algae and may be eaten by fish, birds and humans.

Heliocidaris erythrogramma (a) aboral and (b) oral views.

Animals of rocky shores | 145

Brown sea urchin
Holopneustes porosissimus
Phylum Echinodermata, Class Echinoidea, Family Temnopleuridae

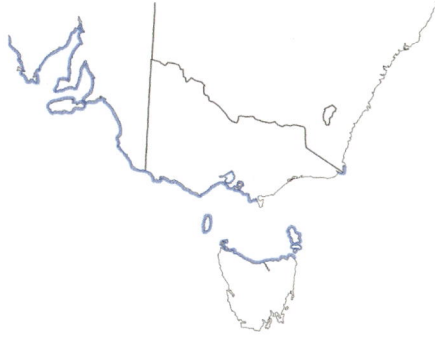

Holopneustes porosissimus (a) aboral and (b) oral views.

Range
WA, SA, Vic., northern Tas.

Appearance
The globular-shaped brown urchin grows to a 'test' diameter of 70 mm and is rarely taller than wide. The short, blunt primary spines are green at the base and red at the tip. The tube feet are purple.

Habitat and ecology
Individual brown urchins live attached to algae in rock pools, shallow channels and subtidally on reefs with moderate wave exposure. Another species of urchin occasionally found intertidally is *Holoptneustes inflatus*, a brown or purple urchin with very long spines.

Sea stars *(Class Asteroidea)*
Biscuit star
Tosia australis
Phylum Echinodermata, Class Asteroidea, Family Goniasteridae

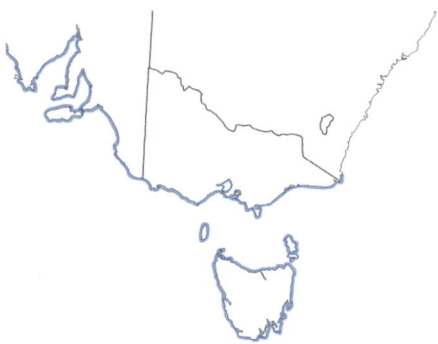

Range
Fremantle, WA, to southern NSW, including Tas.

Appearance
This is a pentagonal-shaped sea star covered with a mosaic of small, flat plates. Six to eight slightly enlarged plates form the margin of each arc between the arm tips. *T. australis* has many colour forms, including brown, pink, red, orange and purple. Individuals grow up to 50 mm across.

Tosia australis examples of colour variation.

Habitat and ecology
This species occurs in the low intertidal zone and subtidally on shores of any type of wave exposure. It feeds on encrusting organisms, such as ascidians, sponges and bryozoans. Gametes are released into the water column and fertilised eggs develop into free-swimming larvae. A recently discovered related species, *T. neossia*, is of similar appearance and has been found mostly in Victoria. The main difference between *T. australis* and *T. neossia* is that the latter broods its young, which develop from larvae that sink to the sea floor and crawl when released from the adult.

Animals of rocky shores | 147

Small green sea star
Parvulastra exigua
Phylum Echinodermata, Class Asteroidea, Family Asterinidae

Parvulastra exigua.

Range
SA, Vic., NSW, Qld, Tas.

Appearance
P. exigua is a small pentagonal sea star that grows to approximately 20 mm across. It has five reduced arms grading into the central disc. The rough upper surface is mottled green to brown.

Habitat and ecology
P. exigua is often found in rock pools, crevices and rock rubble throughout the mid-intertidal zone on shores with moderate wave exposure. It is sometimes found on sand-covered areas of a rocky shore. Individuals feed by everting their stomachs over microscopic plant and animal material. Adults lay egg masses in August and small sea stars emerge very soon after (no planktonic dispersal phase). Two species found only in South Australia, namely *P. vivipara* and *P. parvivipara*, brood their young internally and release them through their body wall.

Eight-armed sea star
Meridiastra calcar
Phylum Echinodermata, Class Asteroidea, Family Asterinidae

Range
Southern WA, SA, Vic., NSW, Qld, Tas.

Appearance
M. calcar has five to nine (but usually eight) distinct but short arms and grows up to 100 mm across. The upper surface colour varies between individuals; it can be any combination of green, blue, orange, red, brown, grey or purple, and is always mottled.

Habitat and ecology
M. calcar is usually found in damp areas or rock pools on shores with low to moderate wave exposure, and can be abundant at any tide level. This animal is omnivorous; it feeds on macroalgae, encrusting organisms and animal remains. In contrast to *P. exigua*, *M. calcar* has planktonic larvae.

Examples of colour variation in *Meridiastra calcar* (photograph a by G. Quinn).

Crimson sea star
Meridiastra gunnii
Phylum Echinodermata, Class Asteroidea, Family Asterinidae

Meridiastra gunnii (a) aboral and (b) oral views.

Range
WA, SA, Vic., NSW, Tas.

Appearance
This sea star has six arms that form a hexagonal shape. It is uniform dark crimson to purple, but often pale around the margin. The tube feet (see Glossary) are orange. The maximum arm radius is 65 mm; more commonly it is 30–40 mm.

Habitat and ecology
The crimson sea star is found under rocks in the lower tidal zone of sheltered reefs and can be abundant. This sea star is omnivorous; it feeds on drift algae, animal remains, compound ascidians and sponges. Gametes are released into the water in late summer to early autumn. There is no larval stage, with fertilised eggs developing and hatching directly into young sea stars.

Spiny sea star, eleven-armed sea star
Coscinasterias muricata
Phylum Echinodermata, Class Asteroidea, Family Asteriidae

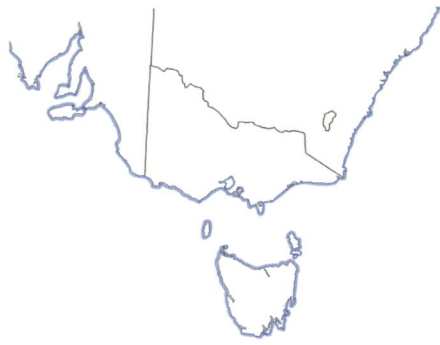

(a) *Costinasterias muricata*. (b) *C. muricata* oral surface showing tube feet.

Range
Australia wide

Appearance
C. muricata is the largest sea star in southern Australia, growing up to 250 mm across. It has rows of spines on its upper surface. The main colour is blue, with markings in brown, red, orange, green, mauve, grey, cream and white. It can have from 7 to 14 arms (typically 11) of varying size on the one animal.

Habitat and ecology
This sea star is found amid reef rubble in the lower intertidal zone, and subtidally on sheltered to moderately exposed shores. It is a predator, feeding on mussels and scallops, and will also scavenge on other material. Although sometimes considered a pest on shellfish farms, **this is a native species** and should not be confused with the voracious introduced pest species *Asterias amurensis* (North Pacific sea star), which is smaller, paler and has only five arms (see p. 151).

Granula sea star
Uniophora granifera
Phylum Echinodermata, Class Asteroidea, Family Asteriidae

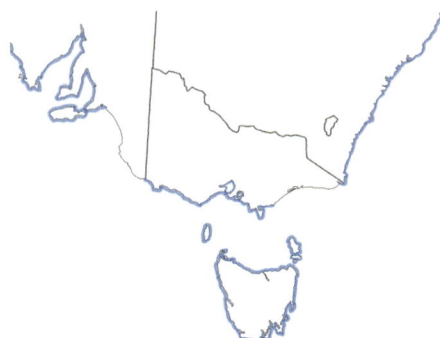

Uniophora granifera (a) aboral and (b) oral surface.

Range
South-west WA, South Australian gulfs to Lismore NSW, Tas.

Appearance
U. granifera is a mottled, rough sea star that can be yellow, pink or purple in colour. The bumps (tubercules) on the aboral surface are often arranged in a zigzag arrangement. *U. granifera* has a similar appearance to the invasive pest *Asterias amurensis*, typically consisting of five arms. The arm tips in *U. granifera* tend to be blunter and are not upturned like *A. amurensis*. *U. granifera* grows to an arm radius of 120 mm.

Habitat and ecology
U. granifera is a predatory sea star that feeds on various small invertebrates, particularly ascidians and bivalve molluscs. This species occurs in boulder field and soft-sediment habitat. Although typically subtidal to depths over 100 m, individuals may occasionally be found in rock pools and the sublittoral fringe zone. It is a native species that is very similar in appearance to the invasive pest species *A. amurensis* (next page), with the main distinguishing feature being the upturned arm tips in *A. amurensis*.

Northern Pacific sea star
Asterias amurensis
Phylum Echinodermata, Class Asteroidea, Family Asteriidae

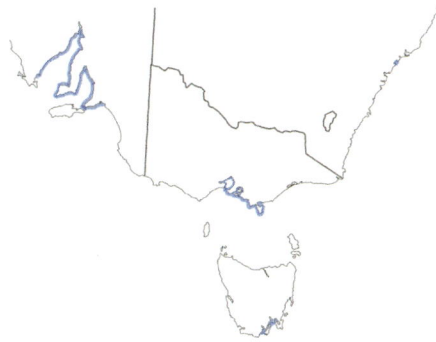

Range
Central SA, Port Phillip Bay, Wilsons Promontory and Gippsland Lakes, Vic., Gosford, NSW, south-eastern Tas.

Appearance
This invasive sea star from the Northern Hemisphere (Japan, Korea, parts of China and Russia) typically has five yellow arms that tend to be upturned at their tips. The arms can also have a purple colouration. This sea star has an arm radius of up to 250 mm (i.e. from the arm tip to the centre of the central disc). *A. amurensis* is easily mistaken for *Uniophora granifera* (previous page) because of similarities in size, the number of arms and colouration, and this may be why it was well established in the Derwent River and Port Phillip Bay before being officially recorded. This species should not be confused with *Coscinasterias muricata* (p. 149), because it has fewer arms and is often much smaller.

Habitat and ecology
Asterias is found on and around pier pylons, rock rubble, seagrass beds and soft sediments. Like other predatory sea stars, it has a varied diet consisting of small invertebrates, particularly bivalve molluscs. It is also an active scavenger and can be attracted to discarded fishing bait. It typically occurs in the subtidal zone to 35 m, but can also be found in the

Asterias amurensis (a) aboral and (b) oral surfaces.

shallow subtidal zone and is often washed up on shore. Northern Pacific sea stars typically spawn between July and October in southern Australian waters and produce a long-lived planktonic larva that can survive in the water column for up to 120 days.

A. amurensis is a pest species in Australian waters. If you see this species, take a photograph and report your find to the relevant fisheries authority in your state.

Ocellate sea star
Nectria ocellata
Phylum Echinodermata, Class Asteroidea, Family Goniasteridae

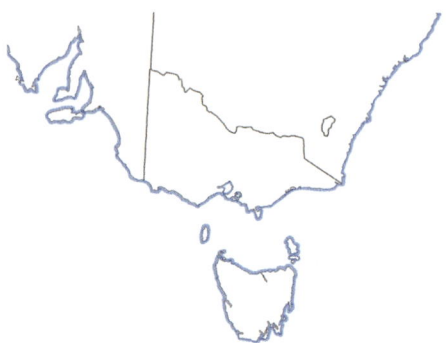

Nectria ocellata.

Range
WA, SA, Vic., NSW, southern Qld, Tas.

Appearance
This sea star is usually bright orange or red and has five arms with a maximum arm radius of 130 mm. The dorsal surface comprises raised plates that become smaller towards the arm tips.

Habitat and ecology
The ocellate sea star is typically a subtidal species, but individuals are sometimes found at the low-tide level on shores with moderate to high wave exposure. *N. ocellata* feeds on sessile invertebrates, such as sponges and ascidians. Sea stars can regenerate missing parts. It is not unusual to find individuals with one or more arms shorter than the rest, and arms lost to predators or accidents are regrown. Some species, such as spiny sea stars (previous page), can regenerate a whole animal from an arm and are able to reproduce by splitting in two, then regenerating the missing parts.

Brittle stars *(Class Ophiuroidea)*
Snake brittle star
Ophionereis schayeri
Phylum Echinodermata, Class Ophiuroidea, Family Ophionereididae

Ophionereis schayeri (a) aboral and (b) oral surfaces.

Range
WA, SA, Vic., NSW, southern Qld, Tas.

Appearance
Brittle stars are similar to sea stars in having body parts arranged in a symmetry of five (pentaradial symmetry) and a water vascular system. They are different in having long, flexible arms and a central disc-shaped body that is clearly distinct from the arms. The arms of brittle stars are solid, whereas those of sea stars are hollow.

O. schayeri grows up to 180 mm from arm tip to arm tip. The light brown central disc is separate from the snake-like arms, which are banded alternately light and dark grey–brown.

Habitat and ecology
O. schayeri is found under rocks or in clumps of organic matter in the lower intertidal zone on shores with moderate to low wave exposure. Individuals move away from the light when the boulders sheltering them are overturned. The snake brittle star feeds on dead plant and animal matter. Unlike sea stars, which crawl using tube feet, brittle stars move by pushing against the ground with their arms. The tube feet are used to selectively feed on fine particulate organic matter (fPOM; see Glossary). The arms will often break off when this animal is handled, so **please don't handle it!**

Sea cucumbers *(Class Holothuroidea)*
Sea cucumber
Lipotrapeza vestiens
Phylum Echinodermata, Class Holothuroidea, Family Phyllophoridae

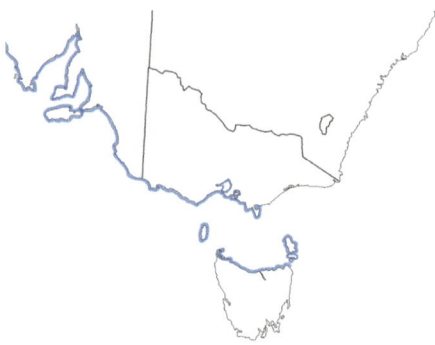

Range
South-west WA, SA, Vic., northern Tas.

Appearance
Sea cucumbers are soft-bodied, sausage-shaped animals with a crown of tentacles (mostly retracted inside the body) at one end. This species can be brown, orange, pink or white. The body surface is usually covered with stones, shell fragments and other debris held in place by sticky tube feet. Individuals can grow to 100 mm long and 40 mm wide.

Habitat and ecology
L. vestiens lives under rocks in the low intertidal zone on shores with low to moderate wave exposure, especially where the substratum is sandy. The oral tentacles are used to gather food particles from the surrounding sand and to filter feed at high tide. Sea cucumbers will often eject their intestines through their mouth (eviscerate) as a defence mechanism if handled or attacked; the intestines are unpalatable to predators.

Lipotrapeza vestiens.

Sea squirts, tunicates
Phylum Chordata

Solitary sea squirts
Tunicates are classified as Chordates because they have a tadpole stage that possesses key chordate features, including a tail, primitive backbone (notochord), dorsal hollow nerve cord and pharyngeal gill slit. Unlike most chordates, such as fish and mammals, tunicates do not develop a backbone. All tunicates are marine.

Cunjevoi, sea squirt
Pyura stolonifera
Phylum Chordata, Subphylum Tunicata, Class Ascidiacea, Family Pyuridae

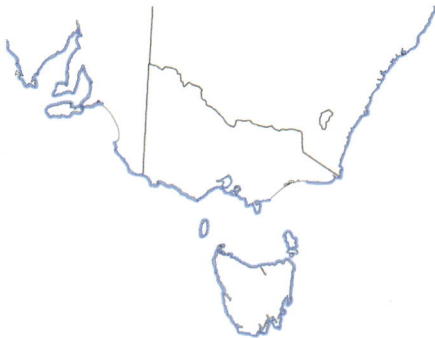

Pyura stolonifera.

Range
Albany, WA, SA, Vic., NSW, southern Qld, Tas.

Appearance
Cunjevoi (say 'cun-je-voy') individuals are roughly cylindrical and grow up to 150 mm high and 50 mm across. Two openings (inhalant and exhalent siphons) lie close together on the upper surface. The outer skin of the animal is hard and leathery, dark red to black and often covered by a dense growth of seaweed.

Habitat and ecology
Cunjevoi live attached to rocks in the lower tide zone on shores with moderate to strong wave exposure. This animal is a filter feeder. It holds water inside its body during low tide, squirting it out through one of its siphons when touched. Eggs and sperm are released into the water and a short-lived tadpole-like larva develops after fertilisation. Cunjevoi are eaten by gastropods and are becoming less abundant due to collection for use as bait.

Southern sea tulip
Pyura australis
Phylum Chordata, Subphylum Tunicata, Class Ascidiacea, Family Pyuridae

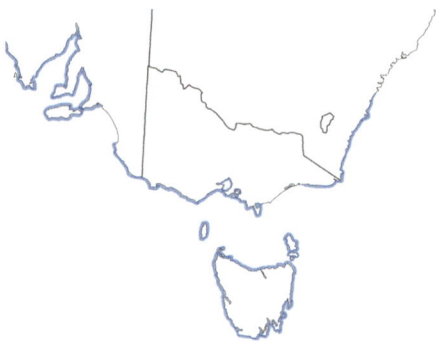

Range
Southern WA, SA, Vic., southern NSW, Tas.

Appearance
P. australis has a leathery red, ovoid body on a stalk that is attached to the substratum. The inhalant and exhalant siphons are not always obvious. The body may have a lumpy appearance due to rows of warty-shaped projections (tubercles).

Habitat and ecology
This sea tulip can be found in rock pools and the subtidal fringe zone, which is only occasionally exposed to the atmosphere during very low (spring) tides. It is more common subtidally. The animal draws in water through an inhalant siphon, filtering out food particles and expelling the water through an exhalant siphon. Gametes are released into the water for external fertilisation, which produces tadpole-like larvae.

Pyura australis.

Compound ascidians
Phylum Chordata, Subphylum Tunicata, Class Ascidiacea, Family Didemnidae

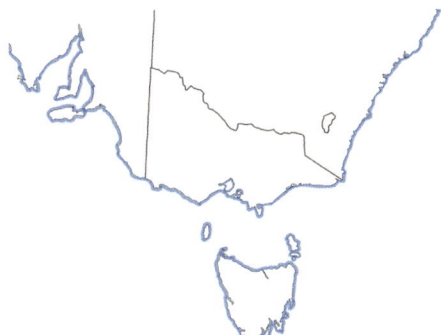

Range
Worldwide

Appearance
Compound ascidians comprise many small individual animals, called zooids, attached together in a gelatinous matrix to form colonies. In the intertidal zone, these compound ascidians usually form flat colonies on the underside of ledges and under rocks. These colonies can be difficult to distinguish from encrusting sponges. Subtidal species can have more lobose growth forms, often with obvious inhalant siphons surrounding a shared exhalant siphon, as in the examples on this page.

Habitat and ecology
Compound ascidians inhabit a diverse range of marine habitats, including intertidal rocky

Two growth forms of compound ascidians.

shores, subtidal rocky and coral reefs, seagrass beds and installed structures, such as pier pylons. They are filter feeders, with the inhalant siphon acting as a pump to suck in water containing microscopic food particles. Reproduction in ascidians is via the release of gametes into the water, with the fertilised eggs developing into non-feeding, free-swimming tadpole larvae for dispersal.

Animals of sandy and muddy shores

Soft-sediment shores vary considerably depending on the wave action and water movement where these habitats occur. Broadly, soft-sediment shores are referred to as either sandy or muddy shores. Sandy shores have coarser particles than muddy shores, with particles ranging from 0.063 to 2.0 mm in diameter. Particles 1.0–2.00 mm in diameter are considered to be coarse sand. Particles larger than 2.0 mm are typically considered to be gravel and pebbles (up to 64 mm), cobbles (64–256 mm) and boulders (>256 mm). In contrast, muddy shores have much finer and stickier sediments that consist of silt, clay and other fine particles that are less than 0.063 mm in diameter. On both shore types, it is still possible to find large particles and shell fragments scattered throughout the sediment matrix at various depths.

Sand is more typical of ocean beaches, whereas fine sediment muddy shores are more typical of sheltered bays and estuaries. In estuaries, the composition of sediment tends to change from coarser sand at the estuary mouth, closer to the ocean, to finer muddy shores further upstream.

Important marine soft-sediment habitats

The marine red, brown and green algae are a conspicuous component of intertidal rocky shores, but not of soft-sediment ecosystems, where the substratum is too unstable for attachment of holdfasts. Instead, three broad groups of flowering plants (angiosperms), namely seagrasses, mangroves and saltmarsh plants, are found in marine soft-sediment environments. The root systems of these plants enable them to inhabit these soft-sediment environments. Where present, macroalgae are often limited to *Caulerpa* spp. (pp. 25–28), with their rhizome-like spreading stolon, and to epiphytes on seagrasses and mangroves.

Seagrasses are predominantly subtidal plants, but some species (e.g. *Zostera muelleri*) may occur in the low intertidal zone. However, seagrasses growing in

Seagrass bed at low tide.

this zone are often damaged by too much UV radiation during low tides, particular during summer. Seagrasses can successfully pollinate underwater, a feature that is unusual in flowering plants. Recent research has shown that pollen is not only carried between plants through water movement, but also on the bodies of small invertebrates.

Mangrove forests, or mangals, can consist of several families and species of trees and shrubs. There are only two species of mangrove that occur in southern Australia (below Sydney): *Aegiceras corniculatum* and *Avicennia marina*. The

Avicennia marina.

Saltmarsh habitat.

grey (or white) mangrove, *Avicennia marina* (p. 57), is the only species that occurs in Victoria, and has its most southerly limit in Corner Inlet. Mangroves occur in the upper intertidal zone and are covered by each high tide. They do not require marine environments, but are better adapted to survive in salty soils than most other terrestrial plants. Many mangroves can also survive in soils with low salinities, but tend to be outcompeted by other terrestrial plant species.

Saltmarsh plants are very tolerant of salty soils and tend to occupy the supralittoral zone, where they are only inundated during the largest (spring) tides. Many saltmarsh plants consist of ground cover species that either have succulent leaves or are grasses that are characterised by narrow, wiry leaves. Living in salty soils is similar to living in a desert, a water-stressed environment. These different plant forms represent two different mechanisms for surviving in a highly water-stressed environment.

Seagrass beds and mangrove forests form important habitat for numerous fish and invertebrates. Living seagrass blades and roots are not typically eaten directly by fish and invertebrates, but they contribute to marine food webs indirectly, in the form of detritus (i.e. as they break down into smaller organic fragments). The seagrass leaves provide cover and refuge for small fish and invertebrates, whereas the complex seagrass rhizome and root complex provides the invertebrates that live in the sediments (infauna) with protection from predators that probe sediments from the surface. Fish that use seagrass beds include commercially important species, such as King George whiting, bream and flathead, and cryptic species such as pipefish, seahorses and clingfish.

Similarly, the complex aerial and prop roots of mangroves (p. 57), together with an array of large trunks that fall into the water, often provide cover for fish and a hard surface for numerous invertebrates and small plants to settle on. Mangrove forests provide protection against coastal erosion and can even reduce the impact of tsunamis.

Life beneath your beach towel

More than 85% of the Australian population lives within 50 km of the coastline, so many Australians spend some time at the beach, with sandy beaches more popular with beach visitors than muddy shores. Sandy beaches often appear devoid of life. However, this is far from true. Many of the animals that inhabit soft-sediment shores remain hidden, burrowed beneath the sediment.

Marine plants and animals often wash onto soft-sediment shores from adjacent reefs, especially after storms; the collective term for these organic materials is 'wrack'. Many beach visitors consider wrack an unsightly, smelly nuisance, resulting in its removal from urban beaches around Australia. However, wrack plays an important role in sandy beach ecosystems. It provides food and shelter for many small invertebrates, fish (when it is washed back into the sea) and shore birds, including protected species, such as the hooded plover (*Thinornis rubricollis*).

There are very few truly marine insects, but insects can be common on sandy beaches because they form the interface between marine and terrestrial ecosystems. There are numerous beetles and flies, in addition to thousands of beach hoppers and pill bugs (crustaceans) that help break down the wrack. Many of these invertebrates (crustaceans, insects and spiders) live beneath the sand or in amongst the wrack. When most people lay down their beach towels, they are often blissfully unaware that they are lying atop of many small 'bugs', including spiders.

Although many of the plants and animals featured in this guide are conspicuous, we felt it important to include some of the not-so-conspicuous soft-sediment fauna that commonly inhabit our beaches and mudflats. Many of these invertebrates are very common (abundant), and although some may be well known to recreational fishers who collect invertebrates for bait, others are cryptic and remain

Seaweed debris (wrack).

largely unnoticed. The colouration of some of these beach animals allows them to blend into the sand very effectively (equivalent to 'sandy beach chameleons').

Typical beach bugs

Amphipods and isopods are two groups of crustaceans that can be very common in marine environments, including sandy beaches. Amphipods (beach hoppers or beach fleas) can be very abundant under the wrack on which they feed. Their spectacular leaps and bounds when disturbed make them obvious to beach goers. When not disturbed, hundreds of tiny holes in the sand surrounding patches of wrack, or in the base of sand dunes, are evidence of their presence. Beach pill bugs (isopods) are less conspicuous; their movements are slow, and they deliberately remain still or roll into a ball when disturbed. Body colour and patterning further aid their camouflage.

A general rule that helps separate the isopods from the amphipods is that amphipods are laterally compressed and tend to lie on their side when not in motion, whereas most isopods are dorsoventrally compressed (flat from back to stomach). The names for the two orders are derived from the direction that the toes point. The toes of isopods all point in the same direction (*iso* = the same), whereas the toes on the front (anterior) and back (posterior) legs of the amphipods point in opposite directions (*amphi* = different). Other common decomposers of wrack include numerous kelp fly species, earwigs and weevils.

Sandy beach predators

Common invertebrate predators on sandy beaches include wolf spiders and rove beetles. Centipedes may also be present. The colouration of wolf spiders helps these hunters to blend in with the sand. Wolf spiders feed on the beetles, flies and crustaceans

(a) A laterally compressed amphipod and (b) a dorsoventrally flattened isopod.

Animals of sandy and muddy shores | 163

living in amongst the wrack. Rove beetles have large mouth parts to seize their prey. They can be quite small (up to around 15 mm long) and resemble an ant. Rove beetles are often black, but can also have different segments that are red or orange.

Like a bug stuck in the mud

In contrast to the dry, sandy beaches that most people are accustomed to, muddy shores contain more moisture due to the fine and closely packed sediment particles, thus creating a very different habitat for organisms. The fine, sticky sediment

Worm castings.

Mictyris burrow.

Macrophthalmus burrow.

Zeacumantus trail.

contains fine particulate organic matter (fPOM; see Glossary), which forms an important food source for many deposit-feeding animals. High amounts of organic matter are also food for certain groups of microbes (bacteria) that consume all the oxygen (i.e. the sediment becomes anoxic) and some groups release a foul-smelling 'rotten egg' gas. Amazingly, some of the larger invertebrates can still live within these anoxic sediments. These invertebrates either have physiological mechanisms to cope with low oxygen or they ventilate their burrows by drawing in oxygenated water from the overlying water column. The movement and burrowing of invertebrates within the sediment can also help oxygenate the sediments. Oxygenated sediment is odourless and maintains its natural colour, whereas anoxic sediment tends to be black and foul smelling. Castings, burrow holes and trails are tell-tale signs of invertebrates living within muddy sediments, with possibly the most well known and obvious being soldier crabs (p. 189).

Segmented or bristle worms
Phylum Annelida, Class Polychaeta

Pod worm, estuarine bait worm, rag worm
Australonereis ehlersi
Phylum Annelida, Class Polychaeta, Family Nereididae

Range
Southern WA to southern Qld and Tas.

Appearance
This infaunal (burrowing) polychaete worm can be found in high densities in sheltered bays and estuaries. Individuals live in a sandy tube, which can sometimes be seen protruding just above the sediment surface. When reproductive, males and females are often a different colour, with males appearing cream/white in colour and females appearing green. When ripe, you may see these worms leave their tubes and shed their gametes into the water. The fertilised eggs hatch into a trochophore larval stage. When fully extended, some individuals can reach the length of your hand (150–200 mm). These worms are often quite distinctive, with long sensory appendages, large eyes and numerous lateral appendages called parapodia ('like legs'). This species is distinguishable from other nereids by having shorter sensory appendages and small, finger-like bumps under the ventral surface.

Australonereis ehlersi.

Habitat and ecology
A. ehlersi is a popular bait species for recreational fishers, who typically use a bait pump to suck worms from the sediment, and is sold as bait in some parts of Victoria. Many nereid worms are active predators and possess a pair of impressive jaws, but this species is thought to be a selective deposit feeder (sifts through the fine sediment for fPOM).

Lugworms, spew worms, blood worms

Phylum Annelida, Class Polychaeta, Family Arenicolidae

Range
Worldwide

Appearance
Lugworms are large, fragile polychaete worms that form J-shaped burrows beneath the sand. Their presence is indicated by a distinctive spaghetti-like faecal casting at the posterior end of their burrow and a small pit at its entrance. These worms grow to around 100 mm in length. They are characterised by a bulbous pharynx, when everted, and fleshy red gills towards the rear (posterior) end.

Habitat and ecology
This species occurs in soft-sediment habitats of intertidal/shallow subtidal areas within sheltered bays and estuaries and along the open coast. It is a non-selective deposit feeder (see pp. 163–164). Lugworms are commonly collected by

Lugworm.

recreational fishers to use as bait and can be extracted from the sand using a bait pump. The reproductive biology of this family can differ between species. Many species are broadcast spawners, meaning that eggs and sperm are released into the water above where fertilisation occurs, whereas some species produce distinctive tear-shaped jelly egg masses that are attached to the sediment. Trochophore larvae develop within these egg masses.

Red thread worms
Phylum Annelida, Class Polychaeta, Family Cirratulidae

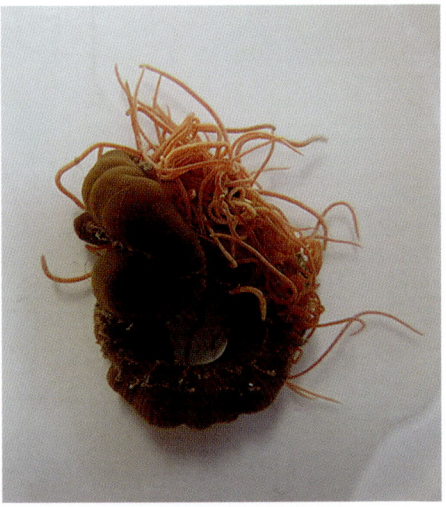

Range
Australia wide

Appearance
These worms can be easily confused with spaghetti worms (Terebellidae) because of the many long filaments (palps) that run along their body. Some of these filaments have fine grooves, which are used for selective deposit feeding; others serve as gills (non-grooved). Red thread worms differ from spaghetti worms in that the filaments extend along the body, whereas they are restricted to the head region in terebellids. Bright orange and red specimens are found along the Victorian coast.

Habitat and ecology
Red thread worms occur in sediment and among or under rocks. Their tentacles may be seen sprawled out across the sediment surface. Red thread worms are selective deposit feeders that gather fPOM from the sediment surface (pp. 163–164). Reproduction within this group is variable; some species are capable of asexual reproduction, but most have separate sexes and exhibit external fertilisation. The reproductive development can also be variable, with both direct development and trochophore larvae occurring within this family.

Red thread worm.

Shells
Phylum Mollusca

Molluscs with coiled shell
(Class Gastropoda)
Rose petal bubble snail
Hydatina physis
Phylum Mollusca, Class Gastropoda, Family Aplustridae

Range
Australia's east coast, north of eastern Victoria; worldwide (tropical and temperate)

Appearance
An attractive gastropod mollusc with a large, pink muscular foot and a light shell that is characterised by many thin brown/black stripes. The shell spire is up to 55 mm long.

Habitat and ecology
This species occurs on soft-sediment habitats, particularly among seagrass beds and some shallow subtidal reef habitats. These molluscs are carnivorous and feed on polychaete worms. They are hermaphroditic and lay white, noodle-like egg masses, similar to those shown for *Aplysia* (p. 198). The eggs hatch within the capsule before being released as a planktonic veliger larva, the characteristic first larval stage of many marine and freshwater molluscs.

Hydatina physis (a) ventral view of shell and (b) dorsal view showing the large muscular foot.

Toothed air breather
Ophicardelus spp.
Phylum Mollusca, Class Gastropoda, Family Ellobiidae

Ophicardelus spp. showing different views and colour variations.

Range
Western Vic., Tas. to northern Qld and parts of NT

Appearance
A small, brown, elongated snail (up to 15 mm) with a characteristic aperture. The inner lip, at the central column (columella), has two distinct folds that resemble teeth.

Habitat and ecology
An air-breathing snail that is typically found on soft sediment along the upper edge of mangroves and in saltmarsh. Members of this family are thought to feed on detritus, algae and other plant matter. The snails belonging to this family are typically hermaphroditic, but descriptions of the reproductive behaviour and development of *Ophicardelus* are difficult to find.

Fragile air breathers
Salinator spp., *Phallomedusa* spp.
Phylum Mollusca, Class Gastropoda, Family Amphibolidae

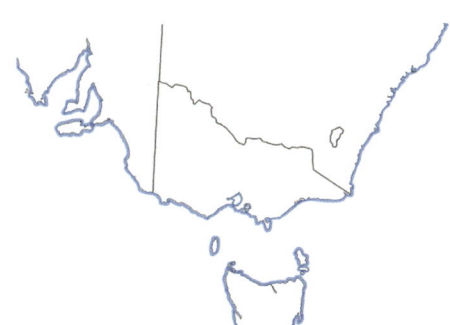

Dorsal and ventral views of *Salinator* sp.

Range
Australia wide

Appearance
A small (up to 10 mm), brown, squat snail with a roundish aperture that is covered by a cap (operculum). These snails have a thin, fragile shell. Individuals can occur in very high numbers in some sheltered bays and estuaries.

Habitat and ecology
An air-breathing, deposit-feeding snail that is often found with *Ophicardelus* along the upper edges of mangroves and in saltmarsh. Members of this group are hermaphroditic and lay string-like egg masses on the sediment surface that become covered in silt. Again, larval behaviour and development in these species does not appear to be well described.

Dog whelks, nassarid gastropods
Nassarius spp.
Phylum Mollusca, Class Gastropoda, Family Nassaridae

Range
Australia wide

Appearance
There are several species of nassarid gastropods that to occur in soft-sediment habitats in south-eastern Australia. The common species grow to a spire length of around 15 mm (typically thumb nail in size). The shell often bears lines of nodules, and some have smooth, polished edges to the aperture; the opening is protected by a cap (operculum). When cruising on the sand, these gastropods have prominent tentacles and a long siphon that protrudes from a distinctive apertural notch.

Habitat and ecology
These snails tend to occupy soft-sediment habitats in intertidal/shallow subtidal areas in sheltered bays and estuaries. Their reproductive biology can vary, but most species produce small egg capsules. Nassarids are active scavengers (carnivores) that can be readily attracted, in large numbers, to discarded bait and dead fish.

(a) *Nassarius pauperatus*, (b) *Nassarius pyrrhus* and (c) *Nassarius burchardi*.

Mud whelks, club whelk, Hercules club whelk, Sydney whelk
Pyrazus ebeninus
Phylum Mollusca, Class Gastropoda, Family Batillaridae

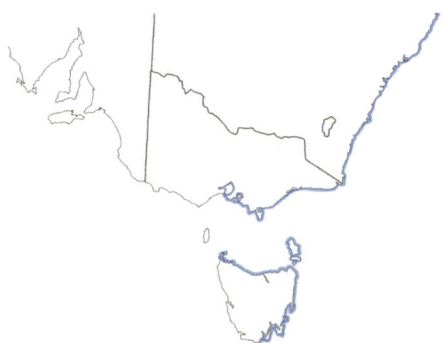

Range
South-eastern Australia to northern Qld

Appearance
A large (up to 97 mm), heavy and conspicuous snail with a solid shell. The shell tends to be dull and brown to grey in colour. Individuals can be found congregating in large numbers on sheltered sand flats and among seagrass. The outer lip of the aperture can be smooth and flared, and the shell consists of numerous pointed nodules.

Habitat and ecology
Although it may look like a veracious, predatory snail, *P. ebeninus* consumes detritus and algae. These snails can occur on sand flats with and without mangroves and seagrass. Significant numbers of shells often appear in Aboriginal middens and were consumed by early settlers. Members of this mollusc family produce egg capsules, with eggs developing into trochophore, and then veliger, larvae that eventually swim out of the capsule. Despite their high abundance in some areas, little is known about the ecology or reproductive biology of Australian batillariids.

Pyrazus ebeninus (a) ventral and (b) dorsal views.

Small mud whelk, mud creepers
Zeacumantus diemenensis
Phylum Mollusca, Class Gastropoda, Family Batillaridae

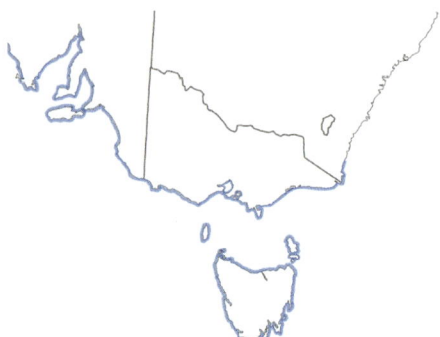

Zeacumantus diemenensis showing variation in form and colour. Operculum visible in (b).

Range
South-west WA to southern NSW, and Tas.

Appearance
The shell of *Z. diemenensis* resembles a mini *Batillaria australis*, varying in colour from brown to grey. The shell of *Z. diemenensis* can grow to a spire height of 35 mm. Like other batillariids, this species can be found in very high numbers on sand flats and seagrass beds in sheltered bays and estuaries.

Habitat and ecology
Individuals can be common on sheltered sand and mudflats, inside and outside of seagrass. Like other members within this family of snails, *Z. diemenensis* is a detritus feeder and is likely to exhibit similar reproductive biology.

Australian mud whelk
Batillaria australis
Phylum Mollusca, Class Gastropoda, Family Batillaridae

Batillaria australis: (a) dorsal (left) and ventral (right) views, and (b) individual with epiphytes growing on shell.

Range
Vic. to northern Qld, but also occurs in parts of WA (likely recent introduction)

Appearance
Another large snail (up to 60 mm) that looks similar to *Pyrazus ebeninus* and occupies similar habitat. However, this species does not grow as large as *P. ebeninus*, has a different aperture and the shell is not as solid and heavy. These large mud whelks are known to carry flatworm parasites that are spread by wading birds.

Habitat and ecology
B. australis has a similar habitat and ecology to *P. ebeninus*.

Moon snail, sand snail
Conuber conicum
Phylum Mollusca, Class Gastropoda, Family Naticidae

Range
Australia wide

Appearance
C. conicum is a common snail found on sand and mudflats of estuaries, bays and open coast, but can be cryptic because it spends time buried beneath the sediment. Individuals can be found by following trails left in the sand. Shell colouration is variable, but shells are often brown/purplish with an orange/brown band on some shell whorls. *C. conicum* a large, fleshy, muscular foot that is protected, when withdrawn, by a distinct brown operculum. Spire height is up to 50 mm.

Habitat and ecology
This snail occupies soft-sediment habitats in intertidal/shallow subtidal areas in sheltered bays and estuaries. These snails are carnivores, and the aperture differs from that of other predatory snails by being round to oval without a prominent siphonal groove. This family of snail is responsible for the neat, parabolic or bevelled bore holes found in clam shells washed up on the shore. These bore holes differ from those created by another group of predatory gastropods, the muricids, which drill holes with straight/parallel sides. Some species of moon snails produce the sausage-shaped jelly egg masses that often wash ashore (p. 198).

Conuber conicum (a) dorsal and (b) ventral view.
(c) *Katelysia* shell with a *Conuber* bore hole.

Molluscs with two shells
(Class Bivalvia)
Brown shell
Hiatula alba
Previously *Soletellina alba*

Phylum Mollusca, Class Bivalvia, Family Psammobiidae

Range
WA, SA, Vic., NSW and Tas.

Appearance
This burrowing (infaunal) bivalve is often brown/purple but can sometimes be yellow or cream in appearance. The brown outer skin (periostracum) on the shell is often conspicuous, particular on empty shells. This species has two very brittle shells that can be easily broken with your fingers. Individuals rarely exceed 45 mm in length. They have two long, white siphons that allow larger individuals to access the overlying water column to feed, even when burrowed up to a depth of 35 cm in the sediment. A large, white, muscular foot allows them to burrow quickly into the sand or mud.

Habitat and ecology
Typically found in subtidal areas, but can be found on intertidal sand/mudflats and in seagrass

Hiatula alba. (a) Siphon holes next to two exposed bivalves. (b) Juvenile specimen.

beds in tidal estuaries. The presence of these bivalves can be identified on the sediment surface by two small, closely spaced holes (one for each siphon). This species can be susceptible to mass mortality events following winter flooding, which is evidenced by hundreds of empty shells washing up on the shore. The feeding ecology of this species remains largely unknown, but other members of this family are deposit feeders that use their inhalant siphons to vacuum food from the bottom. This species is targeted by recreational fishers for bait and has been sold for bait in some parts of Victoria.

Stepped venerid, ridged cockle, ladder venus, sand cockle, enigma venus
Katelysia scalarina
Phylum Mollusca, Class Bivalvia, Family Veneridae

(a) *Katelysia scalarina* lateral view. (b) *K. scalarina* with the brown anemone *Anthopleura hermaphroditica* attached to its shell.

Range
South-west WA to southern NSW, Tas.

Appearance
A small (up to 40 mm), white cockle with a solid, robust shell that possesses distinct concentric ridges. The colouration of the shell can be quite variable, from plain white to the presence of faint zigzag patterns on some individuals. This species may be recognisable to the public as a popular table clam, particularly in spaghetti vongole. *K. scalarina* individuals live just below the sediment surface, where their presence can be given away by attached algae or anemones, or by a pair of siphons.

Habitat and ecology
These clams occur in soft-sediment habitats in intertidal/shallow subtidal areas in sheltered bays and estuaries. This species is a deposit-/suspension-feeding clam that can be very abundant in some parts of Port Phillip Bay.

Members of this family tend to have separate sexes, are broadcast spawners (see Glossary) and produce planktonic trochophore larvae that develop into veliger larvae before leaving the plankton to settle in the sediment. This bivalve forms a symbiotic relationship with the mudflat anemone *Anthopleura hermaphroditica* (p. 67).

Little wing pearl shell
Electroma papilionacea
Phylum Mollusca, Class Bivalvia, Family Pteriidae

Range
SA to north Qld; worldwide in both tropical and temperate waters

Appearance
A small (up to 40 mm), fragile, thin-shelled bivalve that can be very common in some sheltered bays and inlets, forming a dominant component of beach-cast shell deposits. *E. papilionacea* has a mottled brown/green-coloured shell that is often sufficiently thin to be transparent. This species is the only southern (temperate) member of the pearl oyster family.

Habitat and ecology
This bivalve attaches to hard surfaces, including seagrass blades, via byssus threads. *Electroma* is a suspension/filter feeder that consumes fPOM (see Glossary). The pearl oysters (or feather oysters) are very different to the edible oysters that belong to the Family Ostreidae. Possibly the best-known pteriids are the *Pinctada* spp., which are farmed and harvested in northern Australia for their pearls.

(a) *Electroma papilionacea* lateral view. (b) *E. papilionacea* specimen in an aquarium with its extended foot attached to the glass.

Pipi, Goolwa cockle, eugarie
Donax deltoides
Phylum Mollusca, Class Bivalvia, Family Donacidae

Donax deltoides washed up in the swash zone.

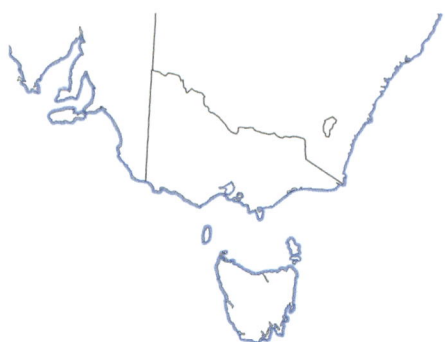

Range
SA to southern Qld, including Tas.

Appearance
D. deltoides is a robust, rapidly burrowing bivalve with a solid, wedge-shaped shell (up to 60 mm). It is a commercially harvested species that is popular as bait for recreational fishers, but is becoming an increasingly popular table clam. Can vary in colour from white to cream, yellow and green with a pink/purple tint. Juveniles can be mistaken for *Paphies* spp.

Habitat and ecology
These popular clams are found in the swash zone of exposed sandy beaches with a wide surf zone. Individuals live just below the sediment surface and can sometimes be seen 'surfing' with the incoming and receding waves. *D. deltoides* can be found in very large abundance on some beaches. These bivalves are filter/suspension feeders that feed on surf phytoplankton. They reproduce by broadcast spawning throughout the spring and summer and produce planktonic larvae that spend from 6 to 8 weeks in the plankton.

Animals of sandy and muddy shores | 179

Elongate little wedge shell, narrow wedge shell, shiny wedge shell
Paphies angusta
Phylum Mollusca, Class Bivalvia, Family Mesodesmatidae

Paphies angusta.

Range
Australia wide, except NT

Appearance
A small (up to 35 mm), white bivalve that looks similar to *Donax deltoides*, but has a smaller maximum size and a more emphasised wedge shape. Individuals can be very common in some sheltered bays and inlets, including some beaches in Port Phillip Bay.

Habitat and ecology
These small clams occur on intertidal sandy beaches in sheltered bays and along the exposed coast. They are often found among *D. deltoides* on exposed coasts but can also occur in sheltered areas without *Donax*. Like other bivalves, this species is a filter/suspension feeder. The reproductive biology of *P. angusta* appears to be understudied. Species from the same genus in New Zealand (the pipi *P. australis* and the tua tua *P. subtriangulata*) spawn during spring and autumn.

Shining theora, fragile semele
Theora sp.
Phylum Mollusca, Class Bivalvia, Family Semelidae

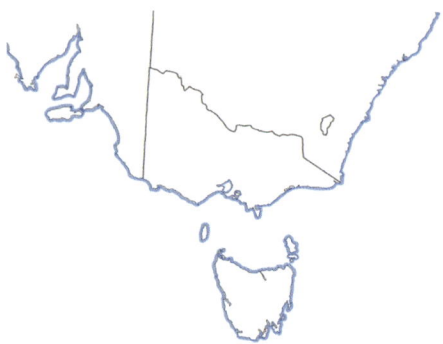

Theora sp.

Range
Australia wide

Appearance
A small, white, transparent bivalve with a very thin, fragile shell. Abundant on sheltered sand flats in Port Phillip Bay, in amongst *Katelysia scalarina* and *Paphies angusta*. *Theora lubrica* is an introduced species to Port Phillip Bay and many parts of south-eastern Australia. Grows to around 20 mm in length.

Habitat and ecology
This fragile clam is an infaunal, deposit-feeding bivalve that lives in sandy sediment to water depths of up to 50 m. Some members of this genus occur in estuarine sediments that have a history of industrial waste discharge (pollution-tolerant species). Individuals of the introduced species *T. lubrica* are fast growing and short lived. Very little ecological information exists for Australian endemic members of genus *Theora* or Family Semelidae.

Animals with jointed limbs
Phylum Arthropoda

Sand hoppers, sea slaters, shrimp and crabs (Class Malacostraca)

Sand hoppers (Order Amphipoda)
Beach hopper, sand flea
Talorchestia sp.
Phylum Arthropoda, Class Malacostraca, Order Amphipoda, Family Talitridae

Talorchestia sp.

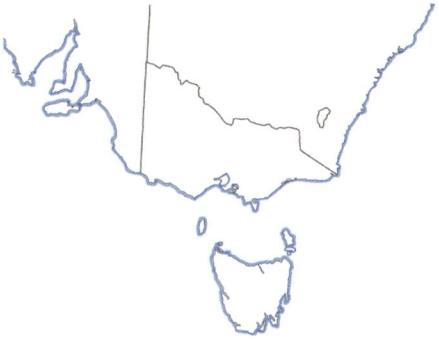

Range
Worldwide

Appearance
These locally abundant invertebrates are typically small (<10 mm), bilaterally compressed, white and brown/orange crustaceans that can dominate assemblages under wrack on beaches with moderate to high wave energy. Their colouration also helps them blend into the sand. They are remarkable jumpers.

Habitat and ecology
There are numerous species and genera of talitrid beach hoppers, which also occur inland. This photograph represents typical individuals found under wrack. They are likely to be an important food source for a variety of shorebirds and offshore fish species. Beach hoppers are important detritivores that contribute to the breakdown of wrack deposits and carrion. Females brood their young in a marsupium on their ventral surface. These animals are collected and used for bait by recreational fishers. Some species are positively phototactic (i.e. attracted to light).

Sea slaters (Order Isopoda)
Pill bug, slater bugs
Actaecia sp.
Phylum Arthropoda, Class Malacostraca, Order Isopoda, Family Scyphacidae

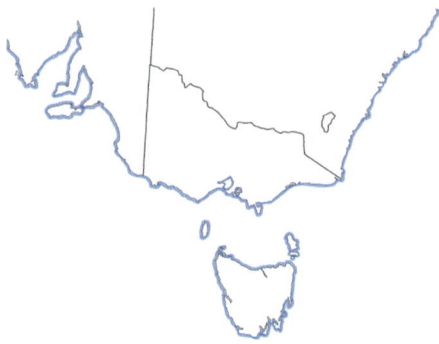

Range
Australia wide

Appearance
These isopods are dorsoventrally flattened, like most other members of this order. They closely resemble species that commonly inhabit backyard gardens, but are distinctly mottled, with brown, black and orange patterns that help them blend in with the sand. These small crustaceans are typically less than 10 mm long, and can roll into a ball when threatened.

Habitat and ecology
Actaecia sp. is another abundant group of crustaceans that often cohabitate beaches with talitrid amphipods under wrack. They also carry their young in a marsupium on the ventral surface. Like the amphipods, *Actaecia* species are important detritivores that feed on wrack and carrion. They cannot jump like the talitrid amphipods, and are slower moving.

Actaecia sp. (a) dorsal and ventral views. (b) Example of an individual rolled up into a ball.

Shrimp and crabs (Order Decapoda)
Ghost shrimp, marine yabby, bass yabby, one-armed bandit, clicker
Trypaea australiensis
Phylum Arthropoda, Class Malacostraca, Family Callianassidae

Trypaea australiensis.

Range
Western Vic. to northern Qld

Appearance
These elegant shrimp have a pink colouration with a slightly transparent carapace. A deep burrowing shrimp that grows up to 65 mm. The entrance of the burrow is characterised by a conical, volcano-shaped mound. *T. australiensis* may be confused with *Filhollianassa ceramica*, but has a large hook on the second segment of the large nipper (chelipeds) and a fringe of dense hairs attached to the first pair of antennae.

Habitat and ecology
T. australiensis is a popular recreational bait species that is collected by a bait pump and is commercially sold in several parts of Australia. It is a delicate, but agile species that can burrow quickly if removed from its burrow.
T. australiensis does not have a hard, robust and/or spiny carapace like other yabbies, shrimp and lobsters. Ghost shrimp are selective deposit feeders. They carry eggs under their tail that develop into long-lived planktonic larvae.

Ghost shrimp, marine yabby, bass yabby, one-armed bandit, clicker
Filhollianassa ceramica
Previously *Biffarius ceramica*

Phylum Arthropoda, Class Malacostraca, Family Callianassidae

Range
Central SA, central Vic., southern NSW, Tas.

Appearance
F. ceramica is a delicate, burrowing shrimp with tinges of red or pink and a transparent carapace that grows up to 90 mm. *F. ceramica* lives in burrows well below the sediment surface. The entrance of the burrow is characterised by a conical, volcano-shaped mound. *F. ceramica* may be confused with *Trypaea australiensis*, but the lower margin of the fourth segment of the pincer in *F. ceramica* is equipped with a serrated tooth and ridge.

Habitat and ecology
Like *Trypaea*, this shrimp is a popular recreational bait species that is collected by a bait pump and is commercially sold in several parts of Australia. It has a very similar external morphology and build to other ghost shrimp species. Ghost shrimp are selective deposit feeders, and their continuous burrowing behaviour and reworking of the sediment plays an important ecological and biogeochemical role.

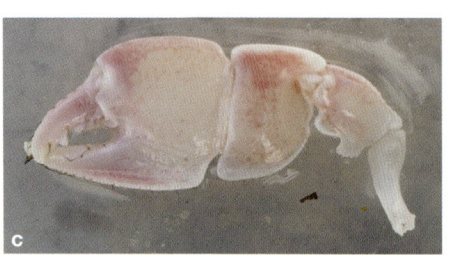

Filhollianassa ceramica (a) dorsal and (b) side views. (c, d) Claws of *F. ceramica* showing serrations on the lower edge of the second segment.

Pistol shrimp, snapping shrimp, clickers
Alpheus richardsoni
Phylum Arthropoda, Class Malacostraca, Family Alpheidae

Alpheus richardsoni.

Range
Central WA to southern Qld, including Tas.

Appearance
This amazing shrimp has the appearance of army camouflage, with khaki stripes. It grows up to 60 mm in length. Pistol shrimp have a large pincer (cheliped) equipped with a peg-and-socket joint that allows them to make a loud clicking sound. One cheliped is larger than the other. The sound generated by the click may be used to stun other unsuspecting invertebrates.

Habitat and ecology
A. richardsoni can be very common inhabitants of soft-sediment environments, on bare mudflats and in amongst seagrass. They either live in burrows or in amongst rubble or under rocks. The sexes are separate. Alpheid shrimp have several larval stages during development from fertilised egg to adult. Some alpheids are carnivorous; others also consume detritus and seagrass.

Smooth pebble crab, nut crabs
Bellidilia laevis
Phylum Arthropoda, Class Malacostraca, Family Leucosiidae

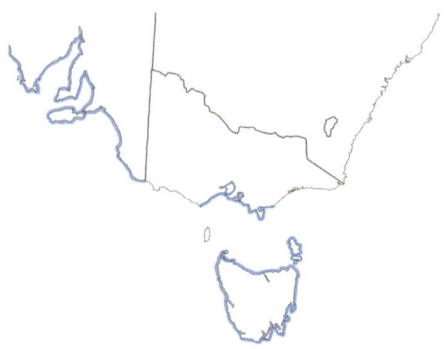

Range
Southern WA, SA, central Vic. and Tas.

Appearance
These crabs have a distinctive, round, smooth carapace that is often light brown/fawn or grey in appearance with four white dots. *B. laevis* is a very active crab when cruising on the sediment surface and can be found in high abundance in seagrass beds. *B. laevis* can quickly burrow into the sand when disturbed. These crabs have very long legs and nippers (chelipeds) relative to their carapace size. They grow to around 25–30 mm carapace width (or the size of a large tom bowler marble).

Habitat and ecology
Pebble crabs can be common in seagrass beds and sandflats on sheltered muddy shores. They are very active and aggressive scavengers and feed on a range of carrion. These crabs produce several larval stages (zoea) and a final megalopa stage that metamorphoses into the fully formed juvenile.

(a) *Bellidilia laevis* anterior view. (b) Ventral view of a male.

Semaphore crab
Heloecius cordiformis
Phylum Arthropoda, Class Malacostraca, Family Heloeciidae

Heloecius cordiformis.

Range
Central Vic. to southern Qld and eastern Tas.

Appearance
H. cordiformis is typically purple, but can be mottled light brown with a cylindrical carapace that is up to 30 mm wide. This species also has distinctly bent claws. The eyes are very long and more closely spaced than those of *Tasmanoplax latifrons*.

Habitat and ecology
H. cordiformis can be locally abundant and digs deep burrows into intertidal mudflats, including mangrove areas. It is a selective deposit feeder that uses it claws to 'spoon' sediment into its mouth parts, which sift organic matter from the sand. Members of this family often have radiating trails of 'sediment balls' from their burrows that are discarded by the crab after feeding.

Southern sentinel crab
Tasmanoplax latifrons
Previously *Macropthalmus latifrons*

Phylum Arthropoda, Class Malacostraca, Family Macrophthalmidae

Range
Gulfs of SA, east of central Vic. to central NSW, eastern Tas.

Appearance
These crabs have a light yellow/brown to purplish carapace up to 25 mm wide. The carapace has two distinct notches on each side, which helps to easily distinguish this species from *Heloecius cardiformis*.

Habitat and ecology
T. latifrons is found in similar habitat to *H. cardiformis* (intertidal muddy shores and mangroves). These crabs are extensive burrowers and selective deposit feeders. *T. latifrons* and *H. cardiformis* are often seen close to the entrance of their burrow, but quickly retreat when approached, so are rarely observed directly by many people.

Tasmanoplax latifrons (a) dorsal and (b) anterior views.

Soldier crab
Mictyris longicarpus
Phylum Arthropoda, Class Malacostraca, Family Mictyridae

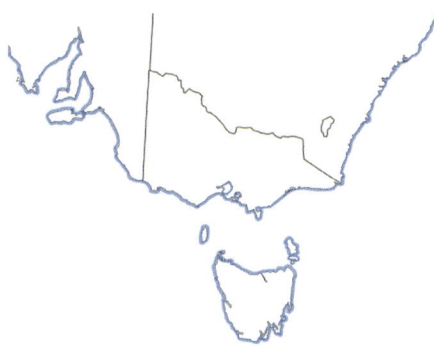

Mictyris longicarpus (a) anterior and (b) dorsal views.

Range
SA, Vic., Tas. and NSW

Appearance
These small, iconic crabs have a spherical carapace up to 20 mm wide that is blue to grey on its dorsal surface, white on the ventral surface and purple/pink on the sides. There are six species in the genus *Mictyris*. Another characteristic feature of all six species of *Mictyris* is that they walk in a forward direction, unlike most other crabs, which move sideways.

Habitat and ecology
The common name, soldier crab, comes from the large 'armies' of individuals that march along intertidal sand and mudflats during low tide. The crabs burrow into the sediment on their side, in a cork-screw fashion, leaving behind a characteristic floral pattern in the sediment. These crabs are selective deposit feeders that filter sand with their multiple mouth parts to strip organic matter from the sediment. As with many crab species, the sexes are separate. The fertilised eggs develop into zoea larvae that go through several developmental stages before being recognisable as juvenile crabs.

Surf crab, sand crab
Ovalipes australiensis
Phylum Arthropoda, Class Malacostraca, Family Ovalipidae

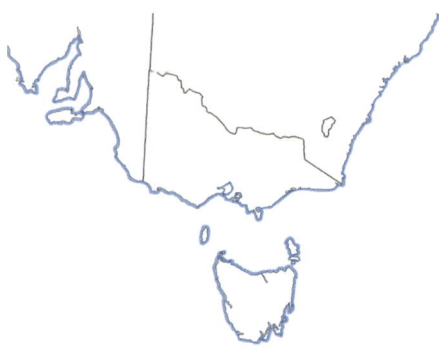

Range
Southern WA to southern Qld and around Tas.

Appearance
O. *australiensis* is a crab with a light grey/fawn-coloured carapace up to 105 mm across. This species has two distinct red dots at the base of the carapace. The most posterior pair of legs is modified into paddle-shaped appendages. The crabs use these legs to swim.

Habitat and ecology
O. *australiensis* can be common in the intertidal zone of sandy shores and occurs subtidally to a depth of 35 m. The members of this family are also active burrowers and often sit just below the sediment surface waiting for unsuspecting prey, occasionally latching onto the toes of waders or paddlers. They can also be a nuisance to recreational fishers because they will scavenge and steal the fishers' bait. In addition to scavenging for food, these crabs are active predators.

Ovalipes australiensis (a) dorsal and (b) ventral (male) views. (c) Carapace washed up on the shore.

Shore crabs

Paragrapsus gaimardii,
speckled shore crab

Paragrapsus laevis,
mottled shore crab

Phylum Arthropoda, Class Malacostraca, Family Varunidae

(a) *Paragrapsus gaimardii* and (b) *Paragrapsus laevis*.

yellow/orange to brown, with numerous black specks on the legs and carapace, and has solid pincers. The carapace can be up to 45 mm wide. *P. laevis* has yellow/cream blotches/mottling on the carapace and legs, with a carapace up to 40 mm wide. These two *Paragrapsus* species can also be separated by observing differences in the tufts of hair that occur on the first pair of walking legs (anterior for *P. laevis*, posterior for *P. gaimardii*).

Habitat and ecology
P. gaimardii is commonly found in sheltered bays and estuaries. Individuals either burrow or live under rocks and other debris. They are omnivorous scavengers that eat rotting algae/plant matter and will not reject discarded fishing bait. *P. laevis* either burrow or occur under rocks on estuarine muddy shores and in seagrass beds. These crabs are carnivorous species that tend to forage at night.

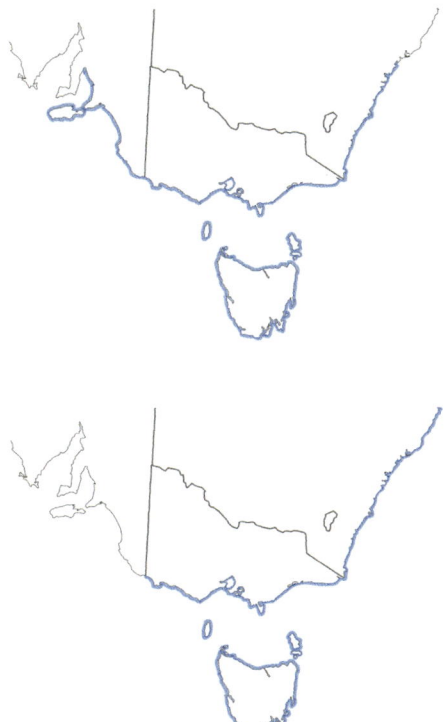

Range for *P. gaimardi* (top) and *P. laevis* (bottom).

Range
P. gaimardii: central SA to southern NSW and Tas.

P. laevis: south-west Vic. to southern Qld, including Tas.

Appearance
P. gaimardii and *P. laevis* both have two notches on each side of the carapace. *P. gaimardii* can be

Burrowing shore crab
Leptograpsodes octodentatus
Phylum Arthropoda, Class Malacostraca, Family Leptograpsodidae

Leptograpsodes octodentatus.

Range
WA, SA, Vic., NSW, and Tas.

Appearance
L. octodentatus is a strong, robust crab with large claws and a circular, green and mottled brown carapace up to 60 mm wide. This species has a single notch on each side of the carapace, just below the eye.

Habitat and ecology
This crab usually burrows above the high-tide mark on the shore. As with several crab species, *L. octodentatus* is more active at night, when it leaves the burrow to forage. *L. octodentatus* is a carnivorous species. The specimen shown in this image was found on King Island.

Millipedes, insects and spiders
Portuguese black millipede
Ommatoiulus moreleti
Phylum Arthropoda, Class Diplopoda, Family Julidae

Ommatoiulus moreleti (a) rolled into a ball and (b) lateral view.

Range
SA to northern NSW and Tas.

Appearance
These are shiny, black millipedes that grow to around 45 mm long. They can be found in very high numbers.

Habitat and ecology
This millipede can be found in various habitats, ranging from sand dunes to sandy beaches, rocky shores and inland. *O. moreleti* is widespread and has no natural predators in Australia. When handled, millipedes can secrete a strong-smelling yellow secretion (a quinone compound). They are known herbivores, and possibly feed on wrack deposits and dune vegetation when inhabiting sandy shores. Female *O. moreleti* lay their eggs in a soil chamber over summer and into early autumn. The eggs hatch and the immature juveniles go through several moults before becoming mature. This millipede is an invasive species that has been introduced to Australia from the Iberian Peninsula.

Kelp beetles

Sphargeris physodes, Scymena amphibia

Phylum Arthropoda, Class Insecta, Family Tenebrionidae

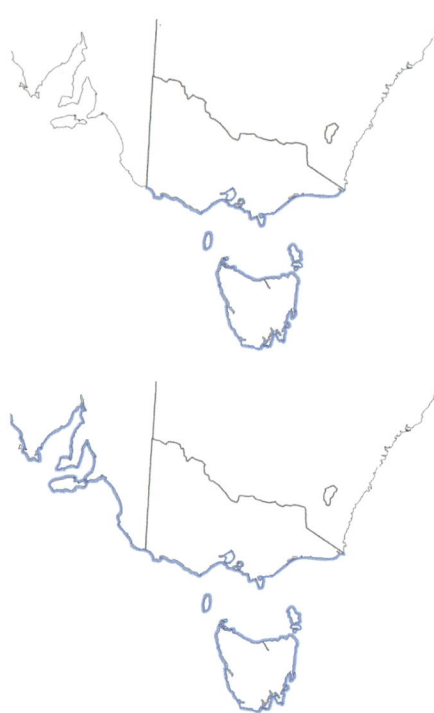

Range
S. physodes: Vic. and Tas.
S. amphibia: SA, Vic. and Tas.

Appearance
Kelp beetles are a golden brown colour, which helps them blend into the sand. These beetles are slightly larger than a lady beetle.

Habitat and ecology
There are several species of beetle that occupy wrack on our sheltered and exposed sandy beaches. Collectively, these are known as kelp beetles or darkling beetles. These beetles are often very cryptic to the public, hidden among the wrack and under the sand. The beetles are also detritivores that help with the breakdown of wrack. The various larval stages of these species look like a 'grub' and are very different to the adult form.

(a) Adult *Sphargeris physodes*, (b) adult *Scymena amphibia* and (c) kelp beetle larva.

Rove beetles
Phylum Arthropoda, Class Insecta, Family Staphylinidae

Range
Worldwide

Appearance
Rove beetles that occur on the sandy beaches of southern Australia tend to be black, elongate insects that have short front wings (elytra), so that you can clearly see the jointed abdomen. Some of the smaller species can look like ants if not closely inspected. The southern Australian beach species tend to be smaller than 15 mm. They are agile beetles and have large, ominous-looking mandibles. They will often quickly fly away once they are disturbed or uncovered in the wrack.

Habitat and ecology
These slender, free-roaming, predatory beetles are often hidden among the wrack. Coastal rove beetles tend to feed on kelp maggots, but are also likely to feed on other small invertebrates and will scavenge on carrion.

Rove beetle without (a) and with (b) wings extended.

Wolf spiders
Phylum Arthropoda, Class Arachnida, Family Lycosidae

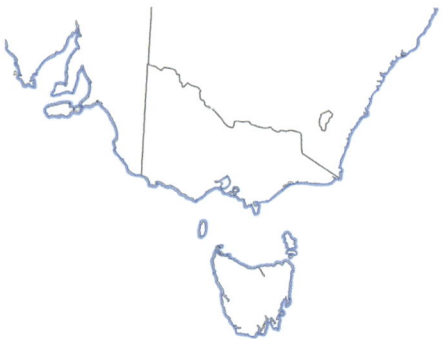

Wolf spider.

Range
Worldwide

Appearance
The sandy beach wolf spiders tend to have a variegated colouration, consisting of brown, black, white and yellow, which helps them blend in with the sand. They can range in size but tend to be no larger a 10-cent piece (leg span).

Habitat and ecology
Wolf spiders are common and occupy nearly all habitat types in Australia, from wet and dry forests to backyard gardens and sandy beaches and dunes. Some dig burrows, but the sandy beach species appear to be free roaming and hide under cover during the day, but may be seen roaming on the sand and among wrack at night. Their eyes can glow/sparkle under torchlight. The females carry their eggs in a round, white, silk 'pillowcase' and will carry the hatched young on their back. Wolf spiders are veracious predators that feed on other ground-dwelling invertebrates, which is likely to include the kelp beetles, small crustaceans and other insects that occupy the wrack.

Egg masses

Most intertidal invertebrates reproduce by releasing eggs and sperm into the water, where external fertilisation takes place. However, some species reproduce by internal fertilisation, depositing the developing embryos into egg masses. Most of the egg masses that you are likely to encounter as a beachgoer will belong to gastropod molluscs (sea snails). This section provides images of some of the common eggs masses that may be found attached to hard surfaces or wash up on the shore.

Lugworm egg mass.

Siphonaria sp. egg mass.

Bedeva vinosa egg mass (photograph by G. Quinn).

Bembicium nanum eggs.

Moon snail egg mass.

Dicathais orbita eggs.

Cominella lineolata with eggs.

Australaria australasia eggs.

Aplysia sp. eggs.

Egg masses | 199

Conus anemone eggs.

Squid eggs.

All washed up: natural marine debris

The natural marine debris you might find when beachcombing can be a guide to what is living in the ocean and on the sea floor.

Seaweeds detached from nearshore reefs are common (*Ecklonia radiata* pictured in image a) and are often in piles called wrack (see p. 161).

By-the-wind sailor *Velella* sp. (b) and the Portuguese man o' war *Physalia* sp. (c) (Phylum Cnidaria) are single animals that consist of a floating colony of zooids (different units) that provide different functions. The colony is stabilised by floating zooid, with *Vellela* bearing an erect, transparent 'sail'. Some of the zooids have a defensive role, some are involved in feeding and others are involved in reproduction. The long tentacles of the man o' war house powerful stinging cells that contain neurotoxins to capture and paralyse prey such as small fish. These stinging cells will be well known to many surfers; even beached, the man o' war can deliver a painful toxic sting.

Ecklonia radiata.

Velella sp.

Numerous individuals of the mosaic or blubber jelly *Catostylus mosaicus* (d) wash ashore, especially in Port Phillip Bay, where they can form large swarms. *C. mosaicus* is a large jelly with a wide bell (up to 350 mm) and is characterised by having eight sets of three trailing arms. Like many cnidarians, the colour of these jellies is typically associated with the presence of zooxanthellae (symbiotic algae).

The paper nautilus *Argonauta nodosa* (e) is an octopus that lives in the ocean surface waters. The female produces a fragile, parchment-like shell to brood her eggs. The male of this species is very small and does not have a shell. Although not a common find, the very beautiful shell of the paper nautilus may be washed up on beaches in mass strandings.

Physalia sp.

Nautilus sp.

Catostylus mosaicus (photograph by P. Davis).

A cuttlebone (f, g) is the internal shell of a cuttlefish and is used to control buoyancy. Cuttlefish are cephalopod molluscs; they are related to octopuses but have two extra arms (tentacles) that are used to catch prey. Cuttlefish are nocturnal predators, preying on a range of invertebrates and fish. It is possible to identify the species from the shape of the cuttlebone (e.g. the cuttlebone in image g is from the giant cuttlefish – see image h).

Cuttlebone of the giant cuttlefish *Sepia apama*.

Stalked barnacles (*Lepas* sp., image i) are often washed ashore attached to bits of wood and other floating objects. They are related to shore barnacles, but differ in that they attach to hard surfaces via a flexible stalk. The body of the animal is surrounded by five hard, white plates.

Live giant cuttlefish *Sepia apama* (photograph by P. Davis).

Cuttlebone of the New Holland cuttlefish.

Lepas sp.

The females of some shark species lay egg cases from which the young sharks hatch. The examples shown here belong to the Port Jackson shark *Heterodontus portusjacksoni* (j) and the endemic swell shark *Cephaloscyllium laticeps* (k), named for its ability to increase its body size by inflating its stomach. The spirals of the former are used to hold the egg case in a rock crevice. Skates (stingray-like relatives of sharks) also lay egg cases; these are commonly called 'mermaid's purses' (l).

Fish carcasses also wash up on shore. Pictured here are the remains of a leatherjacket *Meuschenia* sp. (m) and a weedy (or common) sea dragon *Phyllopteryx taeniolatus* (n). Endemic to southern Australia, the latter is the State marine emblem for Victoria. Sea dragons live among seagrass and seaweeds on moderately wave-exposed reefs. The male carries the developing eggs under its tail.

Swell shark egg case.

Skate egg case.

Leatherjacket remains washed up on shore.

Weedy sea dragon remains washed up on shore.

Port Jackson shark egg case.

Glossary

Antheridia: The structures in which sperm/spermatozoids are produced.
Asexual reproduction: Creation and development of offspring with identical genetics to the parent (i.e. a clone), which can occur via fragmentation, budding or from an unfertilised egg. *see* **parthenogenesis**
Aperture: Typically pertains to the opening in a shell (e.g. of a snail or barnacle).
Aristotle's lantern: Intricate feeding structure of sea urchins that includes the jaws.
Benthic: Occurring on or within the sea floor.
Bilateral symmetry: Two-sided symmetry, where an organism has distinct left and right sides, head and tail ends, and dorsal (back) and ventral (front) surfaces.
Blade: Analogous to the leaf of a plant; the primary site of photosynthesis. Usually a broad, flattened branch of the main plant.
Broadcast spawning: Release of gametes into the water column for fertilisation.
Brooding: Carrying eggs or developing young either within or close to the body.
Budding: An outgrowth from the main body of an animal (the parent) that may detach and develop into offspring with identical genetic material as the parent (i.e. a clone).
Byssus threads: Strong fibres produced by some bivalve molluscs that provide them with a form of non-permanent attachment to a hard surface.
Carapace: The hard 'shell' that covers the thorax of arthropods (e.g. the shell of a crab).
Chaetae: The hairs, or bristles, that are characteristic of marine polychaetes (*poly* = many, *chaetae* = bristles).
Cilia: Hair-like structures that protrude from cells.
Cirri: The feeding appendages of barnacles and the attachment apparatus of feather stars.
Cnidocyte: A specialised cell typically associated with cnidarians that houses a stinging structure. There are several types of stinging structures (e.g.

nematocysts, which are barbed or venomous coiled threads that are fired to capture prey or for self-defence).

Coenocytic: Constructed of a large, single multinucleate cell without dividing cross-walls that usually takes the form of a branching tube or siphon.

Colonial: A group or cluster of individuals that are typically derived asexually from a single colonising individual (e.g. corals, bryozoans and ascidians).

Conceptacle: An internal cavity with a small opening in which the male and/or female reproductive structures (antheridia and oogonia respectively) are situated in some red algae and fucoid brown algae.

Cotyledons: 'Seed leaves'; the first leaves that appear on the stem of a flowering plant seedling.

cPOM: Coarse particulate organic matter. Consists of large pieces of organic material, such as beach-cast algae, plants and other deceased marine and terrestrial organisms.

Deposit feeder: Animals that typically feed on organic matter, particularly fine particulate organic matter (fPOM), that settles on the sea floor.

Detritivore: Animals that feed on non-living organic matter (detritus); i.e. 'mop up the scraps'.

Detritus: Collective term for organic matter in the environment, which includes cPOM and fPOM.

Dioecious: Having male and female reproductive systems in separate individuals. For algae, this means separate male and female thalli producing sperm and eggs respectively.

Dorsal: Pertains to the back of a bilaterally symmetrical animal.

Endemic: Pertains to the distribution of biota; restricted to a particular country or region.

Epifauna: Typically refers to mobile or attached animals that live on hard surfaces, such as rock, sediment or other plants and animals.

Epiphyte: Typically a small plant or alga that grows on a larger host plant or seaweed.

fPOM: Fine particulate organic matter. Small organic particles that originate from algae, plants and dead animals (cPOM). Important food source for detritivores (deposit and suspension feeders) and plays an important role in other biochemical processes.

Frond: The main part of the thallus above the holdfast, including the stipe (where present) and blade. The branching pattern of fronds is often important for species identification.

Gamete: Microscopic reproductive cells that only have a single strand of DNA (haploid) and usually require fertilisation/fusion with another gamete to form a diploid (double-stranded DNA) cell that can germinate/develop.

Gametangia: The structures in which gametes are produced, although in many algae gametes are produced in normal cells rather than in specialised structures.

Gametophytes: Multicellular stage in the life cycle of a seaweed that has reproductive structures known as gametangia that produce gametes. Gametangia that produce sperm are called antheridia and those that produce eggs are called oogonia. Gametes usually require fertilisation to progress to the next stage in the life cycle.

Genicula (singular)/**geniculae** (plural): The flexible, non-calcified joints of coralline red algae.

Girdle: Bristly, scaly or fleshy outer border of tissue of chitons where the eight shell plates sit, the features of which can be used to identify species.

Habitat: The place where animals and plants live.

Haptera: Cylindrical root-like branches of a holdfast (especially in kelps) that intertwine, forming complex microhabitats and strong attachment to hard surfaces.

Hermaphrodites: Organisms that have both male and female reproductive systems in the same individual at some stage during their lifetime (can occur simultaneously or sequentially).

Holdfast: Attachment site for seaweed to a hard substratum, usually constructed of either solid tissue, tangled fine rhizoid filaments or thick, cylindrical intertwining 'root-like' haptera.

Infauna: Animals that live (burrow) beneath the surface of the sea floor.

Introvert: The long, extendable tube of peanut worms that terminates in tentacles and the mouth.

Invertebrates: Animals without backbones.

Lamina: Distinct flattened section of a blade, particularly in kelps.

Larva: Developmental stage of many invertebrates.

Lophophore: Crown of tentacles around the mouth of bryozoans that is used for feeding.

Mantle: Soft outer body layer of molluscs that covers the internal organs, creates a cavity for gills or the respiratory surface and secretes the shell. A thin mantle covering internal organs is also present in barnacles.

Marsupium: A brood pouch that protects the eggs and offspring of a breeding organism.

Megalopa: Larval stage of crustaceans in which legs and abdominal appendages are apparent. Follows an earlier zoea stage.

Monoecious: Hermaphroditic, with both male and female reproductive structures present.

Mucron: A short, sharp-pointed or blunt cap on the tip of utricles in some species of *Codium*.
Nauplius: Early larval stage of a crustacean. Has three paired appendages and an unpaired simple eye.
Omnivore: An animal that has a mixed diet of plants and animals.
Oogonia: The structures in which eggs are produced.
Operculum: Moveable covering or plates that protect the opening of a shell or tube.
Palps: Paired fleshy projections from the head of polychaetes that are typically used in feeding or sensory function. Palps are also associated with some molluscs and arthropods.
Parapodia: Small, often fleshy, lateral appendages ('legs') associated with marine bristle worms. Also a key morphological feature of some molluscs (e.g. sea hares).
Periostracum: A thin outer layer, or skin, found on the shell of several molluscs.
Pharynx: A muscular tube (a 'throat') that extends beyond the mouth of an organism.
Parthenogenesis: Development of an embryo from an unfertilised egg.
Planktonic: Occurring in the water column; movement controlled by currents and tides; either permanent (holoplankton) or temporary (meroplankton).
Planula larva: A free-swimming larval stage of several marine invertebrates.
Pneumatophore: 1. The rod-like aerial root of some species of mangrove trees that becomes exposed during low tides.
2. The gas-filled floats associated with some species of floating cnidarians, such as the Portuguese man o' war and some algae.
Pneumostome: The small hole that opens into the mantle cavity of an air-breathing snail or slug (a pulmonate).
Proboscis: Eversible muscular extension or tube that includes the mouth and is used for trapping or collecting food.
Radula: The feeding organ of many molluscs that resembles a ribbon of hard teeth and can be used to inject toxins (cone shells), for scraping algae from rocks or for tearing tissue from prey or carrion.
Ramuli: Small lateral branches of species of algae that have blades that branch many times.
Receptacles: Blade branches that are covered by conceptacles in some brown algae.
Rhinophores: 'Club-' or 'ear'-like tentacles commonly associated with sea slugs that function as chemosensory organs (taste and smell).
Rhizoid: Root-like structure in seaweeds that is used for attachment and consists of small filaments comprising one to several undifferentiated cells.

Rhizomes: Creeping horizontal stems of vascular plants (including seagrasses) that are usually buried within the sediment.

Rhodoliths: Free-living ball-like nodules formed from encrusting coralline red algae that completely encase small rocks, pebbles or sand.

Rostrum: A 'nose' typically associated with crustaceans. It extends the carapace from beyond the eyes and can create a sharp, serrated spine.

Setae: Similar to polychaete bristles, but differ in structure and belong to arthropods.

Sessile: An organism that is permanently attached (fixed to one place).

Settlement: Typically refers to planktonic dispersal stages (e.g. larvae, spores, zygotes) that leave the plankton and settle on the sea floor or substrata, developing into a mobile or sessile juvenile stage.

Sexual reproduction: Reproduction involving the production of two types of haploid gametes with a single set of chromosomes (e.g. eggs and spermatozoa in animals and some (but not all) plants) that fuse during fertilisation to form diploid zygotes that are genetically different from the parents.

Siphonal groove: A notch situated at the anterior end of the shell aperture in carnivorous snails.

Spawning: Release of gametes, fertilised eggs or spores, usually into the water.

Spicules: A diverse range of microscopic skeletal elements found in sponges.

Spire: Pertains to the spiral structure found within the coiled shell of a snail.

Spore: Microscopic reproductive cells of algae (and some higher plants like mosses and ferns) that are produced in sporangia. Spores may be motile (with little tails called flagella) or non-motile (lack flagella) and can germinate directly (i.e. without fertilisation) into the gametophyte stage.

Sporangia: The structures in which spores are produced.

Sporophytes: The multicellular stage in the life cycle of a seaweed that has reproductive structures known as sporangia that produce spores.

Stipe: Structure analogous to a stem that serves to elevate the blade towards the water surface and the sunlight; usually strong but flexible to resist breakage due to wave action.

Stolon: Creeping horizontal axis from which vertical fronds arise that may sit on the substratum or be buried within the sediment.

Substratum: The base or material on which an organism lives or to which it is attached.

Suspension/filter feeder: An animal that ingests water and filters (or strains) organic particles that are suspended in the overlying water column for food.

Swash zone: Upper part of the beach where waves run up and down the beach slope, causing the sediment to be alternately wetted and drained with each wave.

Symbiotic: A close association between two different organisms that can result in a beneficial, detrimental or neutral relationship for one or either of the two organisms.

Thallus (singular)/**thalli** (plural): The body of a multicellular alga or seaweed.

Trichomes: Hair-like structures in cyanobacteria and some algae.

Trochophore: Larval stage associated with some groups of marine worms and molluscs. Trochophores are spherical or pear-shaped and have a ring of cilia around their middle.

Tube feet: Tube-like extensions occurring in rows on the bodies of sea urchins and sea cucumbers, as well as on the under-surface of the arms of sea stars and brittle stars.

Vascular tissue (plants): Specialised tissues involved in transporting water, nutrients and the sugars produced during photosynthesis throughout the plant.

Veliger: A second larval stage of some marine molluscs that has large ciliated lobes used for swimming, feeding and gas exchange.

Ventral: On or relating to the underside (the belly) of an animal.

Verrucae: 'Wart'-like growths on the column of sea anemones.

Vesicles: Gas-filled sacs or bladders that act as floats in some species of brown algae.

Wrack: Beach-cast material that often consists of large amounts of seaweed (algae) or seagrass and other plant material. Dead animals (carrion) also become entangled in the wrack.

Zoea: One of several larval stages that form part of a crustacean life cycle. Most zoea have a distinctive spine, conspicuous eyes and mouth parts for feeding.

Zooid: The individual animals that make up a colony.

Further reading

Websites

Aboriginal Heritage Tasmania (2020) *Aboriginal Shell Middens*. <https://www.aboriginalheritage.tas.gov.au/cultural-heritage/aboriginal-shell-middens>.
Algaebase. <https://www.algaebase.org>.
Atlas of Living Australia. <https://www.ala.org.au>.
Australian Government (2022) *State of the Environment 2021*. <https://soe.dcceew.gov.au>.
Baldock R (2010–2019) *Identification Factsheets of the Marine Benthic Flora (Algae) of Southern Australia*. eFlora SA. <http://www.flora.sa.gov.au/algae_revealed/index.shtml>.
State Government of Victoria (2021) *Fact Sheet: Aboriginal Coastal Shell Middens*. <https://www.firstpeoplesrelations.vic.gov.au/fact-sheet-aboriginal-coastal-shell-middens>.
CSIRO (2022) *Marine Climate Responses*. <https://www.csiro.au/en/research/natural-environment/oceans/marine-climate-responses>.
iNaturalist. <https://www.inaturalist.org/>.
National Species List (n.d.) *Vascular Plants. Australian Plant Census (APC)*. <https://biodiversity.org.au/nsl/services/search/taxonomy>.
Taxonomic Toolkit for Marine Life of Port Phillip Bay. <https://portphillipmarinelife.net.au/>.
World Register of Marine Species (WoRMS). <http://www.marinespecies.org>.
Womersley HBS (n.d.) *The Marine Benthic Flora of Southern Australia*. <http://www.flora.sa.gov.au/algae_flora/The_Marine_Benthic_Flora_of_SA_static_index.shtml>.

Books

Anderson D (Ed.) (2001) *Invertebrate Zoology*. 2nd edn. Oxford University Press, Oxford.
Bennett I (1992) *Australian Seashores; Adapted From W.J. Dakin's Australian Seashores*. Collins/Angus & Robertson, Sydney.
Breidahl H (1997) *Australia's Southern Shores*. Lothian, Melbourne.
Clayton M, King R (1990) *Biology of Marine Plants*. Longman Cheshire, Melbourne.
Edgar G (2001) *Australian Marine Habitats in Temperate Waters*. Reed New Holland, Sydney.

Edgar G (2008) *Australian Marine Life; The Plants and Animals of Temperate Waters*. Reed New Holland Press, Sydney.

Fitzsimons J, Wescott G (2016) *Big, Bold and Blue: Lessons from Australia's Marine Protected Areas*. CSIRO Publishing, Melbourne.

Fuhrer B, Christianson I, Clayton M, Allender B (1981) *Seaweeds of Australia*. Reed, Sydney.

Gowlett-Holmes K (2008) *A Field Guide to Marine Invertebrates of South Australia*. Notomares, Hobart.

Huisman J (2019) *Marine Plants of Australia*. UWA Publishing, Perth.

Sainty G, Hosking J, Carr G, Adam P (Eds) (2012) *Estuary Plants and What's Happening to Them in South-east Australia*. Sainty Books, Sydney.

Shepherd S, Thomas I (Eds) (1989) *Marine Invertebrates of Southern Australia (Parts I–III)*. South Australian Government Printer, Adelaide.

Index of common names

Abalone 82
Adelaide periwinkle 92–3
Anemones 64–8
Anemone cone 107, 199
Asian shore crab 141
Austral seablite 55
Australian mud whelk 173
Australian saltgrass 55
Australian shore slater 124
Australian tubeworm 75
Axe head mussel 112

Banded goblets 23
Barnacles 116–22
Bass yabby 183, 184
Beach hopper 181
Beaded glasswort 53
Beaded samphire 53
Beaded zigzag weed 45
Beaked mussel 111
Bear seaweed crab 130
Biscuit star 146
Bivalves 110–113
Black crow 96
Black-finger crab 136
Black-mouthed conniwink 97
Blood worms 166
Blubber jelly 201
Blue-green algae 18
Blue-ringed octopus 114
Branching sponge 63
Bristle worms 70–6, 165–7
Brittle stars 153

Brown fan weeds 38
Brown sea urchin 145
Brown shell 175
Bubble weed 39
Bull kelp 40
Burrowing shore crab 192
By-the-wind sailor 200

Cap-shaped false limpet 85
Cartrut shell 102
Checkered topshell 92–3
Chequerboard whelk 104, 198
Chipolata weed 34
Chitons 79–81
Clickers 183–5
Club-leafed zigzag weed 43
Club whelk 171
Common blue mussel 110
Common ear shell 91
Common hermit crab 127
Common reed 56
Common (European) shore crab 133
Common surf barnacle 121
Compound ascidians 157
Conniwinks 97
Coralline red algae 20–1
Cousin Itt weed 37
Crab burrows 163
Crayweed 41
Creeping brookweed 56
Crimson sea star 148

Cunjevoi 155
Cuttlefish 202

Dead-man's fingers 36
Dog whelk 102, 170, 198

Eastern reef crab 136
Eastern shore (rough) barnacle 118
Eelgrass 50
Eight-armed sea star 147–8
Elephant snail 83
Eleven-armed sea star 149
Elongate false ear shell 91
Elongate little wedge shell 179
Encrusting brown algae 34
Encrusting sponge 63
Enigma venus 176
Estuarine bait worm 165
Estuarine barnacle 122
Estuarine coral 75
Estuarine sea spider 131
Estuary shrimp 126
Eugarie 178
European shore crab 133

False crab 127
False limpets 90–1, 197
False spider crabs 131–2
Flat-backed crabs 131–2
Flat-lobed zigzag weed 44
Flatworms 69
Flea mussel 111

Index of common names

Forked zigzag weed 42
Fragile air breathers 169
Fragile semele 180

Garweed 50
Ghost shrimp 183, 184
Giant chiton 79
Giant kelp 48
Glass shrimp 126
Globe algae 35
Golden kelp 47
Golf ball sponge 63
Goolwa cockle 178
Granula sea star 150
Grasses 55–6
Green anemone 65
Green bait weed 32
Green sea velvet 28–9
Grey mangrove 57

Hairy seaweed crab 130
Hairy stalked barnacle 116
Hairy stone crab 128
Herbs 56
Hercules club whelk 171
Honeycomb barnacle 119
Horseshoe mussel 111

Intertidal (ocean beach) slug 89
Iron crab 136

Japanese shore crab 141

Karengo 19
Kelp beetles 194
Kelp lice 125
Keyhole limpet 84

Lace coral 143
Ladder venus 176
Laver 19
Leather kelp 47
Leather tube 34
Lichens 58
Limpets 84–8, 90–1
Limpet paint 34

Lined whelk 104
Little brown mussel 112
Little shore crab 140
Little wing pearl shell 177
Liverwort seaweed 32
Lugworms 166, 197

Maori octopus 115
Marine pill bug 123
Marine yabby 183, 184
Mermaid's necklace 30
Mitre shells 100
Modest four-plated barnacle 122
Moon snail 174, 198
Mud creepers 172
Mud whelks 171–3
Mudflat anemone 67, 176
Mulberry shell 103

Narrow wedge shell 179
Nassarid gastropods 170
Native fan worm 73
Neptune's necklace 39
Nodular periwinkle 98–9
Nori 19
Northern Pacific sea star 151
Notched shore crab 139
Nudibranchs 109
Nut crab 186

Ocean beach slug 89
Ocellate sea star 152
One-armed bandit 183, 184

Paddleweed 51
Painted lady 95
Paper nautilus 201
Peanut worms 78
Periwinkles 98–9
Petterd's limpet 88
Pheasant shell 95
Pied limpet 87
Pill bug 182
Pipi 178
Pistol shrimp 185

Pod worm 165
Port Jackson shark 203
Portuguese black millipede 193
Portuguese man o' war 200–1
Proboscis worms 77
Purple-mouthed rock shell 101, 197
Purple sea urchin 144
Purple shore crab 138
Pygmy mussel 112
Pyramid periwinkle 98–9

Ragworms 71, 165
Ramshorn crab 129
Red bait crab 142
Red bryozoan 143
Red feather weed 24
Red rock crab 142
Red swimmer crab 134–5
Red thread worms 167
Red waratah 64
Ribbed topshell 92–3
Ribbon worms 77
Ridged cockle 176
Rock pool shrimp 126
Rock whelk 105
Rose petal bubble snail 168
Rosette barnacle 120
Rosy barnacle 121
Rove beetles 195
Rough barnacle 118
Rough rock crab 134–5
Rugose slit limpet 85

Samphire 54
Sand cockle 176
Sand crab 190
Sand flea 181
Sand hoppers 181
Sand snail 174
Sargassum weeds 46
Sausage weed 36
Scale worm 70
Sea apples 28–9
Sea asparagus 53, 54
Sea centipede 125

Sea cucumbers 154
Sea hares 108, 198
Sea lettuce 32–3
Sea nymph 52
Sea slaters 123, 182
Sea slugs 108
Sea squirts 155
Sea stars 146–52
Seagrasses 49–52, 158–9
Segmented worms 70–6, 165–7
Semaphore crab 187
Shellgrit anemone 66
Shining theora 180
Shiny wedge shell 179
Shore crabs 133, 138–41, 191
Shore stalked barnacle 116
Short-tailed nudibranch 109
Shrimp 126, 183–5
Shrubby glasswort 54
Skate 203
Slater bugs 182
Small brown anemone 67
Small green sea star 147
Small mud whelk 172
Smooth pebble crab 186
Smooth shore crab 138
Snails 82–107, 168–74
Snake brittle star 153
Snakelock anemone 65

Snapping shrimp 185
Soldier crab 189
South African cladophora 31
Southern (common) flatworm 69
Southern sea tulip 156
Southern sentinel crab 188
Spaghetti worms 72
Speckled anemone 66
Spengler's triton 105
Spew worms 166
Spiny porcelain crab 127–8
Spiny sea star 149
Spirorbid worms 76
Sponges 63
Squid 199
Stalked barnacles 202
Stepped venerid 176
Strap weed 41
Striped-mouth conniwink 97, 198
Surf barnacle 117
Surf crab 190
Swell shark 203
Swift-footed crab 137
Swimmer crabs 134–5
Sydney coral 74
Sydney rock oyster 113
Sydney whelk 171

Tall-ribbed limpet 87

Tapeweed 52
Terebellid worms 72
Three-pronged flat-backed crab 131–2
Toothed air breather 169
Top shells 92–3
Tulip shell 106, 198
Tunicates 155–7
Turban shell 94
Turfing red alga 22
Two-spined crab 132

Urchins 144–5

Variegated limpet 86
Veined rock shell 101
Velvet crab 134–5

Warrener 94
Wavy topshell 92–3
Weedy sea dragon 203
Whelks 101–5, 170 3
White-striped anemone 68
Winkles 92–3
Wolf spiders 196
Worm castings 163

Zebra topshell 92–3
Zigzag weeds 42–5

Index of scientific names

Acanthaceae 57
Actaecia sp. 182
Actinia tenebrosa 64
Actiniidae 64–7
Afrolittorina praetermissa 98
Alismatales 49–52
Alpheidae 185
Alpheus richardsoni 185
Amarinus laevis 131
Amblychilepas nigrita 84
Amphibolidae 169
Amphibolis antarctica 52
Amphipoda 162, 181
Amphiroa spp. 21
Annelida 70–6, 165–7
Anthopleura hermaphroditica 67
Anthothoe albocincta 68
Aplustridae 168
Aplysia spp. 108, 198
Aplysiidae 108
Arachnida 196
Archaeobalanidae 122
Arenicolidae 166, 187
Argonauta nodosa 201
Arthropoda 116–42, 181–96
Ascidiacea 155–7
Ascomycetes 58
Asterias amurensis 151
Asteriidae 149–51
Asterinidae 147–8
Asteroidea 146–52
Aulactinia veratra 65

Australaria australasia 106, 198
Australonereis ehlersi 165
Austrocochlea constricta 92–3
Austrocochlea porcata 92–3
Austrolittorina unifasciata 98
Austrominius modestus 122
Avicennia marina 57

Ballia callitricha 24
Balliaceae 24
Balliales 24
Bangiaceae 19
Bangiales 19
Bangiophyceae 19
Batillaria australis 172–3
Batillariidae 171–3
Bedeva vinosa 101, 197
Bellidilia laevis 186
Bembicium melanostoma 97
Bembicium nanum 97, 198
Bivalvia 110–13, 175–80
Brachidontes rostratus 111
Brachynotus spinosus 140
Bryopsidales 25–9
Bryozoa 143
Buccinidae 104

Cabestana spengleri 105
Callianassidae 183–4
Callyspongia sp. 63
Caloplaca sp. 58
Capreolia implexa 22
Carcinidae 133–5

Carcinus maenas 133
Catomerus polymerus 117
Catophragmidae 117
Catostylus mosaicus 201
Caulerpa brownii 25
Caulerpa cactoides 27
Caulerpa flexilis 26
Caulerpa lentillifera 25
Caulerpa longifolia 26
Caulerpa spp. 25–8
Caulerpa taxifolia 28
Caulerpa trifaria 27
Cellana tramoserica 86
Cephalopoda 114–15
Cephaloscyllium laticeps 203
Ceramiaceae 23
Ceramiales 23
Ceramium flaccidum 23
Ceratosoma brevicaudatum 109
Chaetomorpha coliformis 30
Chamaesipho tasmanica 119
Chenopodiaceae 53–5
Chlorodiloma adelaidae 92–3
Chlorodiloma odontis 92–3
Chlorophyta 25–33
Chordata 155–7
Chromodorididae 109
Chthamalidae 118–19
Chthamalus antennatus 118
Cirratulidae 167
Cladophora prolifera 31
Cladophoraceae 30–1

Cladophorales 30-2
Cnidaria 64-8
Codium fragile 28-9
Codium mamillosum 28-9
Codium pomoides 28-9
Colpomenia claytoniae 35
Colpomenia ecuticulata 35
Colpomenia peregrina 35
Colpomenia sinuosa 35
Cominella lineolata 104, 198
Conidae 107
Conuber conicum 174
Conus anemone 107
Corallina officinalis 20
Corallinaceae 20
Corallinales 20
Coscinasterias muricata 149
Crustacea 116-42
Cyanobacteria 18
Cyanophyceae 18
Cyclograpsus audouinii 138
Cyclograpsus granulosus 138
Cymatiidae 105
Cymodoceaceae 52
Cystophora moniliformis 45
Cystophora platylobium 44
Cystophora retorta 42
Cystophora spp. 42-5
Cystophora torulosa 43

Demospongiae 63
Dendrilla rosea 63
Dicathais orbita 102, 198
Dictyosphaeria sericea 32
Dictyotaceae 38
Dictyotales 38
Didemnidae 157
Diloma concamerata 92-3
Diogenidae 127
Diplopoda 193
Distichlis distichophylla 56
Donacidae 178
Donax deltoides 178
Durvillaea amatheiae 41
Durvillaea potatorum 41
Durvilleaceae 40

Echinodermata 144-54
Echinoidea 144-5
Echinometridae 144
Ecklonia radiata 47
Ectocarpales 34-5
Electroma papilionacea 177
Ellobiidae 169
Eriphiidae 136
Euidotea bakeri 125
Eupolymnia koorangia 72

Fasciolariidae 106
Ficopomatus enigmaticus 75
Filhollianassa ceramica 184
Fissurellidae 83-5
Florideophyceae 20-4
Fucales 39-46

Galeolaria caespitosa 74
Gastropoda 82-109, 168-74
Gelidiaceae 22
Gelidiales 22
Goniasteridae 146, 152
Grapsidae 137
Guinusia chabrus 142
Gymnolaemata 143

Halicarcinus ovatus 131-2
Haliotidae 82
Haliotis rubra 82
Haliotis scalaris 82
Haliptilon sp. 20
Halophila australis 51
Halophila ovalis 51
Halopteris paniculata 37
Hapalochlaena maculosa 114
Heliocidaris erythrogramma 144
Heloeciidae 137
Heloecius cordiformis 187-8
Hemigrapsus sanguineus 141
Heterodontus portusjacksoni 203
Hiatula alba 175
Hipponyx conicus 94
Holopneustes porosissimus 145

Holothuroidea 154
Hormosira banksii 39
Hormosiraceae 39
Hydatina physis 168
Hydrocharitaceae 51
Hymenosomatidae 131-2

Ibla quadrivalvis 116
Iblidae 116
Insecta 194-5
Iodeteidae 125
Isara carbonaria 100
Ischnochiton australis 80-1
Ischnochiton cariosus 80-1
Ischnochiton contractus 80-1
Ischnochiton elongatus 80-1
Ischnochiton versicolor 80
Ischnochitonidae 80-1
Isopoda 123-5, 162, 182

Jania sp. 20
Julidae 193

Katelysia scalarina 67, 174, 176, 180

Lamiales 57
Laminariaceae 48
Laminariales 47-8
Leathesia difformis 35
Lepas sp. 202
Lepidonotus melanogrammus 70
Leptograpsodes octodentatus 192
Leptograpsodidae 192
Leptograpsus variegatus 137
Lessoniaceae 47
Leucosiidae 186
Lichina sp. 58
Ligia australiensis 124
Ligiidae 124
Liliopsida 49-52
Lipotrapeza vestiens 154
Lithophyllaceae 21
Lithothamnin sp. 20

Index of scientific names

Litocheira bispinosa 132
Litocheiridae 132
Littorinidae 97–9
Lobophora spp. 38
Lomis hirta 128
Lomisidae 128
Lottidae 87–8
Lunella undulata 94
Lycosidae 196

Macroctopus maorum 115
Macrocystis pyrifera 48
Macrophthalmidae 188
Magnoliophyta 49–7
Magnoliopsida 57
Majidae 129–30
Malacostraca 123–42, 181–92
Maxillipoda 116–22
Meridiastra calcar 147–8
Meridiastra gunnii 148
Mesodesmatidae 179
Metagoniolithon stelliferum 21
Meuschenia sp. 203
Mictyridae 189
Mictyris longicarpus 163, 189
Mitridae 100
Mollusca 79–115, 168–80
Montfortula rugosa 85
Mopalidae 79
Mucropetraliella ellerii 143
Muricidae 101–3
Mytilidae 110–12
Mytilus galloprovincialis 110
Mytilus planulatus 110

Nacellidae 86
Nassaridae 170
Nassarius burchardi 170
Nassarius pauperatus 170
Nassarius pyrrhus 170
Nassarius spp. 170
Naticidae 174
Naxia aries 129
Nectocarcinus integrifrons 134–5
Nectocarcinus tuberculosus 134–5

Nectria ocellata 152
Nemertina 77
Nereididae 71, 165
Nerita atramentos 96
Nerita melanotragus 96
Neritidae 96
Nodilittorina pyramidali 98
Nostocales 18
Notheia anomala 39
Notoacmea petterdi 88
Notomithrax ursus 130
Notoplana australis 69
Notoplanidae 69

Ochrophyta 34–48
Octopodidae 115–16
Ommatoiulus moreleti 193
Onchidella nigricans 89
Ophicardelus spp. 169
Ophionereididae 153
Ophionereis schayeri 153
Ophiuroidea 153
Ostreidae 113
Oulactis muscosa 66
Ovalipes australiensis 190
Ovalipidae 190
Ozius truncatus 136

Padina fraseri 38
Paguristes frontalis 127
Palaemon dolospinus 126
Palaemon serenus 126
Palaemonidae 126
Paphies angusta 179
Paragrapsus gaimardii 191
Paragrapsus laevis 191
Paragrapsus quadridentatus 139
Paridotea ungulata 125
Parvulastra exigua 147
Parvulastra parvivipara 147
Parvulastra vivipara 147
Patelloida alticostata 87
Patelloida latistrigata 87
Perenereis spp. 71
Petraliellidae 143

Petrocheles australiensis 127
Phaeophyceae 34–48
Phallomedusa spp. 169
Phascolosoma spp. 78
Phasianella australis 95
Phasianellidae 95
Phragmites australis 56
Phyllophoridae 154
Phyllopteryx taeniolatus 203
Phyllospora comosa 41
Phyllotricha spp. 46
Physalia sp. 201
Pinnotheres pisum 110
Plagusiidae 142
Platyhelminthes 69
Plaxiphora albida 79
Poaceae 55–6
Polychaeta 70–6, 165–7
Polynoidae 70
Polyplacophora 79–81
Porcellanidae 127
Porifera 63
Porolithaceae 21
Porolithon sp. 20
Porphyra spp. 19
Posidonia spp. 52
Posidoniaceae 52
Primulaceae 56
Psammobiidae 175
Pseudoralfsia verrucosa 34
Pseudoralfsiaceae 34
Pteriidae 177
Pyrazus ebeninus 171
Pyropia spp. 19
Pyura australis 156
Pyura stolonifera 155
Pyuridae 155–6

Ralfsiales 34
Rhodophya 19–24
Rivularia firma 18

Sabellastarte australiensis 73
Sabellidae 73
Saccostrea glomerata 113
Sagartiidae 68

Salinator spp. 169
Samolus repens 56
Sarcocornia blackiana 53
Sarcocornia quinqueflora 53
Sargassaceae 42–6
Sargassum spp. 46
Scutus antipodes 83
Scymena amphibia 194
Scyphacidae 182
Scytosiphon lomentaria 34
Scytosiphonaceae 34–5
Scytothamnales 36
Seirococcaceae 41
Semelidae 180
Sepia apama 202
Serpulidae 74–5
Siphonaria diemenensis 90–1, 197
Siphonaria funiculata 90–1
Siphonaria tasmanica 90–1
Siphonaria zelandica 90–1
Siphonariidae 90–1
Siphonocladaceae 32
Sipuncula 78
Sphaceriales 37
Sphaeromatidae 123

Sphargeris physodes 194
Spirorbidae 76
Spirorbis sp. 76
Splachnidiaceae 36
Splachnidium rugosum 36
Staphylinidae 195
Stomatella impertusa 91
Stypocaulaceae 37
Suaeda australis 55

Talitridae 181
Talorchestia sp. 181
Tasmanoplax latifrons 187–8
Tecticornia arbuscula 54
Temnopleuridae 145
Tenebrionidae 194
Tenguella marginalba 103
Tesseropora rosea 121
Tethya sp. 63
Tetraclitella purpurascens 120
Tetraclitidae 120–1
Theora sp. 180
Tosia australis 146
Trochidae 91–3
Trypaea australiensis 183
Tunicata 155–7

Turbellaria 69
Turbinidae 94

Ulva compressa 33
Ulva spp. 32–3
Ulvaceae 32–3
Ulvales 32–3
Ulvophyceae 25–33
Uniophora granifera 150

Varunidae 138–41, 191
Velella sp. 200
Veneridae 176
Verrucaria sp. 58

Xenostrobus pulex 111
Xenostrobus securis 112

Zeacumantus diemenensis 172
Zonaria spp. 38
Zostera muelleri 50
Zostera nigracaulis 50
Zostera polychlamys 50
Zostera tasmanica 50
Zosteraceae 50
Zuzara sp. 123